HYDROINFORMATICS AS SOCIOTECHNOLOGY: PROMOTING INDIVIDUAL STAKEHOLDER PARTICIPATION BY USING NETWORK DISTRIBUTED DECISION SUPPORT SYSTEMS

To Klimentina, Jon and Ilin

Hydroinformatics as Sociotechnology: Promoting Individual Stakeholder Participation by Using Network Distributed Decision Support Systems

DISSERTATION
Submitted in fulfilment of the requirements of
the Board for Doctorates of Delft University of Technology
and of the Academic Board of the International Institute for Infrastructural, Hydraulic
and Environmental Engineering for the Degree of DOCTOR
to be defended in public
on Thursday, 28 February 2002 at 13:30 hours in Delft, The Netherlands

by

ANDREJA JONOSKI
born in Kumanovo, Republic of Macedonia
Master of Science with Distinction, IHE

This dissertation has been approved by the promoter:
Prof. dr. M.B. Abbott TU Delft/IHE Delft, The Netherlands

Members of the Awarding Committee:

Chairman	Rector Magnificus TU Delft, The Netherlands
Co-chairman	Rector IHE Delft, The Netherlands
Prof. B. Petry	IHE Delft, The Netherlands
Prof. dr. ir. W.A.H. Thissen	TU Delft, The Netherlands
Prof. dr. ir. P. van der Veer	TU Delft, The Netherlands
Prof. dr. J.P. O'Kane	University College Cork, Ireland
Prof. D.P. Loucks	Cornell University, USA

ISBN 90 5410 427 9

Contents

Acknowledgements

This study has been entirely supported by the International Institute for Infrastructural, Hydraulic and Environmental Engineering – IHE, in Delft, The Netherlands. I would like to express my sincere gratitude to everybody at IHE who made this work possible. In particular, I would like to thank IHE for fully supporting the unconventional research line followed in this study. There are not many places in the world where such openness and flexibility is offered, and I hope that IHE will continue to be a place where many more future researchers will be able to experience similar freedom for their creative thought.

During the five years spent on this work, I was granted the unique opportunity to work with my promoter Prof. Michael B. Abbott. My deepest gratitude goes to him for his patience, guidance, and continuous faith in my abilities, even when I was facing insurmountable mental and emotional deadlocks. He revealed a whole new world for me, which thoroughly transformed my professional life as hydroinformatician, but above all, it elevated me as a person to heights I never imagined I could attain.

I am indebted to all my colleagues from the hydroinformatics group of IHE for their steady support during my study. My friend and colleague Slavco Velickov, a wizard in front of the computer and an excellent hydroinformatician, gave me indispensable support in the development of *Aquavoice*. Vincent Guinot, the person who knows even the most hidden secrets of numerical models, patiently offered me his understanding of all the intricacies of the MIKE SHE modelling system. During two most critical periods of my work, at the very beginning and at the very end, I was supported by my colleagues, Dimitri Solomatine and Arnold Lobbrecht, respectively. In all the years that we have shared our PhD experience, my friend Yonas Dibike helped me more than he is aware of, with his calmness and clear reasoning.

I am extremely grateful to Jan Luijendijk and Roland Price from the Department of Hydroinformatics and Knowledge Management of IHE for all their support, and especially for their understanding and patience during the last year of my study.

I would like to thank all the members of the examining committee for reading the thesis. In particular, Prof. Bela Petry, Prof. Wil A.H. Thissen, and Prof. Pete D.P. Loucks provided some very valuable comments which contributed to the improvement of the quality of the thesis.

This book is a gift to my parents. To my mother Mirjana, who first taught me both about the beauty of science, and about its limits, and, to my father Jordan, from whom I learned that the struggle for justice doesn't have to be always coloured in red.

Chapter 1

Introduction

About three years ago, while this work was without any sense of definite direction and I was bewildered by the body of extensive and extremely diversified 'background' material that had been given to me, in a short period of only one week I had the pleasure of meeting two old friends. They had both, like myself, left Macedonia several years before and were busy searching for their happiness elsewhere in this world. The first, who had left for North America, and had been living there for almost ten years, came to visit me during a short business trip in Europe. During the couple of all-night-long conversations flooded with memories of our younger days, he also showed an interest in my study and asked me what it is that I am working on for my doctorate. As things were not clear even to myself, I struggled to find some simple words for describing the topic:

"You know…it has something to do with using the Internet for helping ordinary people to express their opinions, and their voices being heard when they are concerned about some water or environmental problems…Some people think that there is great potential in this, and that Internet will radically change our understanding and practising of democracy, and that it is worth making some efforts in this direction…but…, I don't know … perhaps it is just one more of those things overloaded with false hopes and wrong expectations, perhaps it is just the excitement with the *new*…sometimes I get so discouraged, you know … ".

"But why?", he replied "Don't you think that it will be great to be able to say all these things that we were discussing in a pub back home over empty bottles of beer and a half-asleep barman as our only listener? This technology will certainly save many young people from that frustrating feeling that you have so much to say, but there is nobody to listen, except a couple of drinking buddies who carry the same frustrating feeling anyway".

Although I wanted to tell him how his description does not really reflect my study subject, I gave up, and enjoyed his sincere enthusiasm. For a couple of days after he left, I felt much nicer about my work. Then I accidentally met my other friend, living in Western Europe for six years already, and, during our chat, the same question came up. After I had again tried to explain my work in a similarly messy way, he said:

"The Internet, hee?! The latest instrument for controlling *all* of us in *every* possible way... It's the Americans, you know... they hooked us all on e-mail and on the Web so that they can know every possible detail about us...what we think, what we like, what we do...soon they will be trading with our souls...But good luck! I'm happy for you, ...people who know about these things are really needed nowadays...you'll probably get a good job...". His good wishes were nice, but the despair of his thinking shook my two-day-old rediscovered optimism.

I picked these episodes here, among the many more moments of emotional turbulence and intellectual deadlock which followed me through this work, because they are glimpses of two basically opposite perceptions of the Internet which coexist and develop in parallel even today. In fact, they are not limited to the perception of the Information and Communication Technologies (ICT, today brought to presence mainly through the phenomenon of the Internet) alone, but to technology in general. Internet is just one of the many technologies which characterise our societies. It is its novelty and pervasiveness which make it so central at the present time. The fact is, however, that our 'techno-society' as a whole is nowadays viewed in two, radically opposed, ways. In their extreme form, these views may shortly be stated like this:

From one side, the 'pro-technology' stance, as we may call it, perceives our 'techno-society' as the ultimate perfection of our social organisation which allows us to live healthier, wealthier, longer and happier lives than ever before. The 'counter-technology' stance, on the other side, perceives this same stream of invention as the most monstrous construction ever created and indeed one which threatens to destroy humanity and our planet. It is not surprising that those who are maintaining the first outlook are mainly people involved actively in the various developments in the natural sciences and technology, while those on the opposite side are mainly people from the social sciences, the humanities, or active representatives of various religions. Sometimes the attitude looks as though 'it goes with the job' as the saying goes.

It took a lot of time and effort to find myself discovering that what really matters is not which one of the above perceptions is closer to the truth, and that it is even less important to try to position the present study in one camp or the other. What is important is to see what is both below and beyond these perceptions. It will be obvious that this study is indeed about technology, but it is not a *technical* study. It is about society and the special role of technology in society, but it is not a *social* study of technology. The point of departure for this study is the realisation that the problems we are facing today are never only social or only technical, but always *sociotechnical*. These two domains have become so interrelated and mutually interdependent that they are inseparable (Callon, 1986; Law, 1991; Latour, 1987, 1991, 1993). They are inseparable *in reality*, and we have reached the stage where even their conceptual separation as two distinct domains becomes untenable, or at least ceases to be in any way helpful at all.

The consequences of this realisation are twofold. Firstly, we are at the point where our *understanding* becomes limited and even impossible if we maintain our perception of society and technology as two separated domains. Our societies are maintained and critically dependent on technology. Neglecting the role and power of technology will leave us blind in understanding our societies. On the other hand, all our technological advancements are conditioned upon the natures of our social organisations, and failing to recognise this will leave us incapable of understanding technology. Secondly, this understanding becomes critical for determining our *decisions about actions* which we undertake. It forms the basis for making *use* of technological means for resolving social deadlocks, which is the central topic of this study.

While identifying the sociotechnical nature of our present condition as our point of departure, it will be still necessary to travel some way before reaching that point. The very fact that by tradition we are used to talking about society and technology as two separate entities has its deeper origins in the most significant transformation of human thought in our history, during the period known as the 'modern era'. Therefore, in order to grasp the absolute necessity of adopting a sociotechnological approach, in the following chapter (Chapter 2) we will spend some time on the analysis and revision of some particular aspects of the modern era, which will be immediately connected to our point of departure. Most of our attention will be concentrated on a specific period of the modern era, which covers approximately the last three hundred years of human history, identified with the project of the 'Enlightenment' and the subsequent waves of its critiques. An attempt will be made to show how the eruptive and unprecedented transformations (both positive and negative) which have occurred during this period, have their roots in the collision of a Medieval Christian philosophy of other-worldly renunciation with the philosophy of Rational Humanism. These transformations in effect shaped the meanings of 'society' and 'technology' as we most commonly understand them today. The most fundamental transformations occurred in our understanding of the nature of knowledge and the questions of morality and ethics which underlie this understanding.

The analysis will be carried further into our present societies, which are frequently described as post-industrial societies, and our contemporary culture, commonly described as a postmodern culture. It is an era where we look with a critical eye on the project of 'modernity' which again brings about a new understanding of the nature of knowledge, and the new 'postmodern ethics' (Bauman, 1996). We shall try to show that this new understanding is actually crucial in accepting sociotechnolgy as a unified concept. Of great importance here is the contrast between the teleological *modernistic* position, which celebrates the "end of history" (Fukuyama, 1992) in the ideal of liberal democracy and free markets, and the desperate condition of most third world societies, whose members have little reason for celebration. The water- and environment-related problems in these societies are particularly vulnerable to the traditional approaches of "development aid", resulting in the common 'failure' of projects, sometimes with disastrous consequences, and in some cases even amounting to criminal acts (Abbott, 2001; Abbott and Jonoski 2001). An attempt will be made to show how the sociotechnical approach is suited to these societies as well, with special emphasis on the future role of information networks in the development of these societies through the development of *people* (Woods, 1993). We shall have to confront the view expressed so precisely long ago by Heidegger, where *modern* technology is the highest stage of misrepresentation of the essence of being human.

After reaching our point of departure, the focus will be on two specific technologies, which will bring us to our immediate subject of social problems originating from, or related to water and the environment, and the means of resolving these problems with the aid of new kinds of technical means. These two technologies are the Internet (and ICT in general), and Hydroinformatics. Although some practitioners may find these two technologies overlapping, and their separation inadequate, the distinction is absolutely essential. ICT is an overarching phenomenon that affects virtually every aspect of our professional and private lives. However, although its social impact is overwhelming and has to be taken into account, from a hydroinformatics point of view it needs to be regarded only as one 'enabling technology', or technology which has to be put to work for a more specific use, namely the appropriate management of water and the water-dependent natural environment. Hydroinformatics, on the other hand, is defined essentially as a technology, but its social dimension has been recognised from its very beginning (Abbott, 1991, 1996). In fact, the development of hydroinformatics changes its nature in a direction where it is no longer sufficient to talk only about its social dimensions, but possibly to redefine it as a sociotechnical discipline (Abbott, 1999a, 2000).

The ways in which ITC can be used for building new kinds of hydroinformatics systems for decision support based on a massive, direct participation of all the interested parties, including the general public, will be introduced in Chapter 3 of this book. The name given to these new kinds of systems is Network Distributed Decision Support Systems (NDDSSs). These systems are envisaged in terms of the postmodern stance of the changing nature of knowledge, which may be described by two key positions. Firstly, the recognition of the fact that knowledge is increasingly produced, distributed and consumed as a *commodity* (Lyotard 1979//1984), whereby the dominant, and apparently the only legitimising criterion of any knowledge becomes the criterion of *performativity*. Secondly, the re-discovery of the power and relevance of so-called *narrative* knowledge (sometimes the terms 'custom' knowledge, 'traditional' knowledge or 'social' knowledge are used), which during the whole project of the Enlightenment has been continuously disparaged in comparison with *scientific* knowledge. The fear that this is a kind of epistemological relativism, which necessarily leads to an ethical relativism with all its horrible prospects, has been to a large extent overcome, and shown to be unsubstantiated. It is argued that it is rather precisely the neglect of narrative knowledge that so often leads to 'failures' of projects involving 'knowledge-management' components.

The concept of the NDDSS is based on this new understanding of knowledge, and its functional requirements stem from this general position (Jonoski and Abbott, 1998). Conceptually divided into a (scientific) knowledge centre and users' periphery, it aims at mobilising both kinds of knowledge in the processes of decision making. The main tasks of these systems then become the transmissions of scientific knowledge from a 'centre' (which is the primary repository of scientific knowledge) to the users' periphery, and of the narrative knowledge from the users' periphery to the centre. In order to achieve the mobilisation of both kinds of knowledge for the purposes of decision making, it becomes necessary to concentrate not only on the simple transmissions of knowledge (which may be viewed as a purely technical task). Far more important become the appropriate *translations* of knowledge, from the discourse of scientific knowledge into the discourse of narrative knowledge and vice versa. These translations appear then as central problems in the design and development of NDDSSs. It is proposed here that in order to achieve adequate transmission and translation of knowledge in both directions and its mobilisation for decision making purposes, these systems will have to be composed of three basic functional components: 1. Fact engine component, located in the knowledge centre, with the primary role of gathering, organising, providing and distributing relevant scientific knowledge as scientific facts; 2. Judgement engine component, located at the user periphery, with the primary role of combining the scientific facts with the beliefs and attitudes (traditionally represented as some kinds of 'interests' and 'intentions') of the users/participants in the decision making process, in order to provide judgements of that particular user or user group about proposed intentions or interventions; 3. A platform for negotiation and collaboration where, by using appropriate aggregation procedures, the diverse judgements are mobilised for negotiation and collaboration purposes, so as to provide for the *emergence* of commonly accepted, equitable courses of action, with appropriate compensatory arrangements. The key quality that needs to be introduced into these systems throughout the processes of knowledge transmission and translation, is the quality of *transparency* (Abbott and Jonoski, 1998).

One of the key aspects of the NDDSSs, will be the inclusion of individuals and individual interests in the processes of decision making together with the (traditionally dominant) corporate interests of various kinds. The traditional (modernistic) approaches of ascribing absolute dominance of scientific knowledge in the processes of decision making has led to the neglect of the individual interests. The proven power of the 'scientific method' has been frequently detached from its subject matter and applied for purposes which have in fact nothing to do with science at all. Examples of this misuse of science, or of scientific rhetoric, for *oppressing* individuals are abundant, and it is more and more frequently argued that they are increasingly prevalent (e.g. Roy, 1999). On the other hand, although we are speaking here about the need for embracing and mobilising narrative knowledge in these processes, it needs to be understood that this knowledge can in no way be imposed as supra-individual *communal* knowledge. The main aim with these systems is to eliminate as much as possible any kind of obstacles for individual expression. We are not seeking to *confront* the scientific knowledge with this communal knowledge, but to allow individuals to express their concerns, interests and intentions within their own social contexts and in relation to the established scientific facts about the objective reality.

Chapter 4 of this book will present a description of a prototype of an NDDSS, which has been developed within the framework of this study (Jonoski, 1999, 2000). The design and development of this prototype has been based on the requirements identified in Chapter 2 and 3 for an NDDSS, although, within the broader context of the previous chapters this application still lacks many important aspects. This prototype, named *Aquavoice,* has a fact engine component, a judgement engine component and some facilities for negotiation and collaboration. It has been developed for an imaginary case study, which deals with the problems of lowering groundwater tables due to increasing abstractions of groundwater for water supply. A number of conflicting interests arise, as represented by four different institutions, and by any number of individuals who may participate in the decision-forming process. *Aquavoice* provides a dynamic environment for judging different scenarios of groundwater abstractions by each user/participant, according to his or her own interests. These different scenarios can be evaluated by each user of the system individually. The individual evaluations are then also aggregated in order to present the overall preference relations of the whole community. This aggregation, together with the transparent positions of each participant in *Aquavoice,* represents the basic 'social landscape' of the community and may be used to initiate processes of negotiation and collaboration between the participating parties. In effect, *Aquavoice* provides a means of accumulating experience as a basic foundation to the making of judgements (e.g. Husserl, 1948//1973)

At the core of *Aquavoice* is a sophisticated hydrological model of the area in question, developed by using the *MIKE SHE* modelling system. This model acts as the fact engine for *Aquavoice*. Being a fully integrated, physically based, modelling system for the simulation of the whole land phase of the hydrological cycle, *MIKE SHE* is particularly suited for being at the core of a system like *Aquavoice*. Much of the technical work, however had to be expected in order to deploy a *MIKE SHE* model on the Internet. The development of a complete new user interface, adapted to the needs of *Aquavoice* and operating in a browser environment, was a first requirement. Further to this, an Internet-based communication link from this user interface with the core *MIKE SHE* model had to be developed.

For the judgement engine component of *Aquavoice* some methods from the field of multi criteria analysis (MCA) have been adapted. In particular, algorithms for fuzzy multiple-attribute decision making have been implemented in order to obtain the judgements and the evaluations of the users/participants. It has to be noted, however, that these methods are not used here for providing definitive results for preferred courses of action, or even as recommendations to some 'decision maker'. Instead, they are used here as tools for establishing the initial response of the community to the proposed scenarios, serving as a starting point for processes of negotiation and various collaborations. The processes of collaboration and negotiations can then proceed through a chat-like interface which is provided in *Aquavoice*, and any changes of judgements or evaluations which result from these processes are immediately and transparently reflected to all the users of the system.

The main technology used for the development of the *Aquavoice* prototype is Java. This powerful language/platform technology was shown to be particularly useful for development of this kind of Internet applications. On the client side, Java applets are used in combination with the scripting language JavaScript and standard HTML (hypertext mark-up language), while on the server side combinations of the so-called servlet and socket connections are used for achieving adequate communication between the different components of the system. In the view expressed here, there is no other technology which addresses the entire complexity of Internet-based applications more successfully than Java, although there are many other technologies which are sometimes superior to Java in some segments, or for addressing some isolated aspects of Internet and Web-based applications.

Aquavoice, of course has to be considered only as a small first step in the development of NDDSSs. Chapter 5 addresses some of the challenges that lie ahead in the development of NDDSSs. Both in the conceptual design and in all of the three functional components of a NDDSS there are many aspects which will need very much further research. In particular, the potential of so-called 'software agent technology' for future developments in NDDSS is addressed (see, originally, Jonoski and Abbott, 1998). Although it appeared promising at the beginning of this study, this technology has yet to prove its potential for building open-ended Internet-based applications as NDDSSs. The paradigm of 'agent orientation' is, however, recognised to be of great relevance, not only for the development of these systems but also for hydroinformatics in general. This chapter tries to relate the studies of software agents to the design and development of NDDSS.

The concluding chapter of this work summarises all the aspects of NDDSSs design and development. It revises the need for such systems, the approaches that can be followed in their design and development and some possible ways forward. Even though we are just at the beginning with the development of NDDSSs, one question, which needs to be addressed immediately, is the introduction of these systems into our societies. This study attempts to provide a rather complete justification of the need for introducing of these new kinds of systems and to give directions for some ways of going forward in their design and development. However, it leaves open the question of how these systems, as they are currently developed, can be introduced into our societies. In Chapter 6 we briefly investigate what are the possibilities for bringing systems like the NDDSS into practice. It should become rather obvious that, even if they are developed to a high level of sophistication, these systems will meet many obstacles in making their way into a broader acceptance. By their nature, these systems are aimed at reversing the established (top-down) order of decision making, which means that they are also aiming at transforming and even inverting, the established order of knowledge/power relations in most societies. For this reason alone, it is essential to find ways to show that these new systems, which carry with themselves new ways of decision making, are indeed beneficial.

Three possible approaches for the introduction of NDDSSs into practice are proposed. The first (so-called 'edutainment': *edu*cation+enter*tainment* approach) proposes further developments of systems similar to *Aquavoice* as environmental computer-games, in order to raise people's awareness of the possibilities of these systems. The second approach argues for the possibility of the introduction of these systems through actions of governmental and non-governmental agencies, primarily at sub-national and local levels. The third approach investigates the possibility of introducing these systems through the market directly. Each of these three approaches will have its own advantages and disadvantages. At this moment of writing, it is far from clear which approach will be the more likely to succeed in any given social context and, even more importantly, it has to be realised that in different societies and cultures, different approaches will be required.

It is worthwhile to come back, at the end of this introduction, to one point, which hopefully will be much clearer when it will be re-stated in the concluding chapter. While being a study of technology, society, water and environment, this is in fact really a study about people. It is grounded in the faith and the strong conviction of the central role of we humans in shaping a better future for ourselves and for our planet. As for the writer and the most likely readers of this thesis, the water specialists and hydroinformaticians, I hope that it will serve as a reminder that it is our *human* responsibility to put technology at the service to humanity through these new kinds of tools, rather than placing humanity at the service, or even at the doubtful mercy, of the technology.

Chapter 2

Rediscovering sociotechnology

They didn't search for the meaning of life, because their lives were full of meaning.

F.M. Dostoevsky

2.1 Sociotechnology - a unified concept

To say that the nature of the problems we are facing today is sociotechnical appears, at first glance, like stating the obvious. In the fields of water and the environment, it would seem that this sociotechnical nature of problems needs even less elaboration. Our ability to exploit and control the waters of our planet by using continuously improving technological means, has brought immense improvements to our quality of life, and radically transformed the ways in which our societies are organised. On the other hand, this same technological advance, which so conditions our lives in many other areas, is commonly blamed for the continuous destruction of the natural environment, in which water plays a central and indeed irreplaceable role. There is a continuous need for new kinds of social arrangements (the introduction of *policies* that reflect environmental awareness, for example) aimed at controlling and hopefully reversing the adverse effects of 'developmental growth'– such as is crucially and ultimately conditioned by technology.

In fact, all discussions about problems related to development and the environment are immediately entangled within a web of social and technical aspects and social and technical solutions. On a superficial level, such as within the various local or global fora where these issues are brought up for discussion, it seems that a realisation is already surfacing that the social and the technical are, after all, scarcely separable. When it comes to real action, however, this realisation collides with one, apparently impenetrable, mental and organisational structure, reflected and institutionalised in virtually all our professional, educational, economic and social institutions. It is a structure based on the traditional conviction that our world is best understood, and consequently most properly *managed,* if we separate it into two domains, namely the *social* and the *technical.*

One of the central problems in the present study is the problem of *decision making* in the fields of water and the environment, in which fields the technical and the social mix and interact in a multitude of ways. This work is aimed at the elaboration of new kinds of means which, although in themselves technological, will improve the processes of decision making within society as a whole. Of course, this task has already long been tackled under the topic of the design and development of a considerable variety of decision support systems (DSSs). However, one of the most striking aspects of current decision making processes and the development of tools and systems for their support is that they are all still carried out and elaborated within a structure based on the existence of the social and the technical as two separate domains. The processes of decision making are consequently carried out, or analysed, either as processes occurring exclusively within the social domain or as tasks which are to be resolved with purely technical means. The two realms of the social and the technical are of course allowed to call upon each other whenever it is necessary to resolve paradoxes and deadlocks in one or the other, taken separately. But, the separation is maintained and cherished as something useful, and even as an arrangement that is extremely valuable.

In this chapter we shall put forward the proposition that this common, and traditional, understanding of the social and the technical has ceased to have both its original relevance and its original purpose (which latter was, presumably, a better understanding of the totality of both our surrounding world and our own human nature). In fact we would like to propose that this separation increasingly becomes a crucial obstacle (and sometimes even an insurmountable barrier) in the resolution of many of our current problems, and especially in the fields of water and the environment. That this separation is untenable in view of our human condition is not a result of a new discovery by a brilliant mind. It comes out of necessity. The scales of the problems (both temporal and spatial) that we are facing today will not allow us to keep the two realms separated anymore. It will not even allow us to talk about the *influences* of one realm upon another, whenever it suits our analyses. We have reached the point when *sociotechnology* needs to become one unified concept, and probably the only one that promises to offer some solutions to our current problems.

The processes of decision making will therefore have to be approached and analysed within this much more inclusive concept of *sociotechnology.* This chapter is intended to show that following this sociotechnical approach will in fact reveal the need for some radical changes in the practices of decision making, in some sense reversing the established order upon which these processes are traditionally based. This analysis in turn points the way towards a new approach to decision making and, inseparably from this, the need to develop new kinds of tools, such as will be explained later in this work. Thus, although the term 'sociotechnology as a unified concept' seems somewhat obvious and natural when dealing with our current water and environment- related problems, it has to be realised that the approach that is followed here challenges many of the established 'orders' in our present societies. In order to arrive at that realisation it has to be understood that, even while the separation of the realms of the social and the technical is far from being accidental, neither is it born out of human existential necessity. It has to be shown in the first place that this separation is *ideological* in the deepest sense of this word, and that the underlying ideology is one that carries the name of *modernity*.

In what follows in this chapter we shall then try to show how this ideology has sparked processes that are unprecedented in history, and how these processes have brought us into a condition of critical questioning of this very same ideology of modernity. In view of the need for understanding the sociotechnical context, as the proper and the only appropriate locus for situating this work, this inquiry becomes necessary. The modern era, marked by the introduction of the two separate domains of *society* and *nature*, such as were discerned before the modern era as composing one unified and indivisible existence proceeding under Divine providence, has held sway over only a quite short period in the history of mankind, but it has brought eruptive changes. Our present condition reveals to us many of the drawbacks brought about by this separation, and indeed it leads us to the *rediscovery* of a different kind of unity between us and the world.

The name *sociotechnology*, which is given to this rediscovered unity, remains both an unfortunate and a fortunate one. It is unfortunate because the simple collation of the two words describing the two different domains is hardly suitable to define a unified concept, but on the other hand it is fortunate because it reminds us of the present reality that the two concepts are currently separated. Many authors who have intended to introduce the unified concept of sociotechnology (see, for example Latour, 1991), have tried to introduce new descriptors such as: "hybrids", "networks", "actors in networks" etc., but it is obviously very difficult to find appropriate expressions in a vocabulary which has itself been shaped within the context of a separation between nature and the 'technology led' activities of human society.

Even as we are introducing the concept of sociotechnology, as a fusion of the social and the technical, there is one aspect that characterises the present condition in most of our societies. That is the fact that the technical is increasingly seen as far more powerful and determining as compared to the social. As technology penetrates into every single aspect of our lives, serious doubts are promulgated about the prospects for a world *dominated* by technology. This stance is frequently described as *technological determinism* and, being one of the most prevalent in our present times, it will be chosen as a starting point here for a rediscovery of sociotechnology. Its origin will be then traced to a specific period in the modern era, known as *the Enlightenment*. Through an examination of various aspects of the current postmodern critique, we shall then try to reassess our current condition, and see how the unified concept of sociotechnology is actually self-imposing as the only concept that will allow us to deal with our complex contemporary reality. Finally, the extreme relevance of this concept to problems of water and the environment will be presented, and it will be shown how its actual introduction into these fields becomes both a justification for and a central task of *hydroinformatics*.

Since we are starting out her in such a complicated situation, the analysis of which is often difficult, so that the reader may well come to ask "where is all this leading us?", it will be useful to introduce certain 'Theses', by way of marking out our position. These will also serve as references in later chapters, where the 'Theses' will be seen to justify decisions concerning new technologies and social arrangements. We may here formulate the first of these Theses as follows:

Thesis 1: *We are aiming at technologies and social arrangements, as exemplified later in this work, which must be viewed from an integral sociotechnical point of view and cannot be treated as simple combinations of social and technical 'aspects'. One of the principle purposes of these new constructs is to lead us to the (re)discovery of a new kind of unity between us, as individuals, and our worlds of human society and nature.*

2.2 Technology unleashed

In an era when technological advancement plays such a central and dominant role in virtually all aspects of our lives, it becomes increasingly difficult to develop a measured view on our relation with technology. One of the difficulties certainly stems from the fact that this relation has a dual and, at the same time, somewhat paradoxical nature. On one hand technology seems to be omnipresent. Although traditionally conceived as an activity concerned with the domain of the inanimate world of objects and primarily human-made artefacts, the proliferation of these very same artefacts has brought technology into every other domain of our life experiences. Even domains where traditionally there was no recognised role for inanimate objects have been penetrated by technology. The questions of social organisations, relations between the social and the individual, the ways in which we define ourselves and others, and everything else which can be seen as falling under the one rubric of 'relations among humans' only, are all increasingly perceived as *determined* to a greater or lesser degree by technology.

At the same time, although technology seems so omnipresent, it is more and more seen as an *independent* development, being one which escapes our control, and even our understanding. For most 'ordinary' people most technological developments are hardly understandable at all. Even for technologists themselves, the world of technology has become so all-inclusive that it is impossible to understand it in its totality, or predict its pace and direction of development. Technology seems to be everywhere, yet impossible to grasp.

The extreme view that technology has become the critical factor that conditions all our activities, drives all the processes of social change and restructuring, and determines all future developments is known as *technological determinism*. And indeed it does appear as though the case for the proponents of this view is becoming very strong in most of our present societies. With the current spectacular technological advances, especially in the fields of Information and Communication technologies (ICT), even the apparently most 'exclusively human' activities are becoming implicitly determined by technology.

Technology cannot be considered any more just as an assembly of tools, which extends our *physical* capabilities. In the era of the television, the computer and the Internet, technology increasingly occupies the realms traditionally 'reserved' for the activities of the human mind. Work, entertainment, politics, economy, art; it is almost impossible to locate any domain which remains outside of the grip of technology. With the latest advances in ICT and biotechnology even the 'last frontier' which was considered 'untouchable' by technology, that of human life itself, seems to have become an object for technological manipulation. The extreme stance of techno-evolutionism (which is an outgrowth of technological determinism) already predicates an inevitable synergy between these two technologies which will result in species of higher order as compared with us humans. Thus the inevitability of our 'handing over' our world to our 'cyborg' descendants has moved out of science fiction, to be now presented as some sort of evolutionary necessity.

Accepting the stance of technological determinism means that this situation has to be regarded simply as 'a fact of life', to be neither celebrated nor criticised. Outside of this narrow position, however, discussions on the role of technology are spread all the way between the extreme opinions of a 'technological nightmare' and a 'technological paradise'. Even the acolytes of the philosophy of technology are split into two groups (Mitcham, 1994). The so-called 'engineering philosophy of technology' praises technological successes to such an extent as to take technological thought and action as a model for all other human activities. On the other hand, the so-called 'humanities philosophy of technology' insists that technological thought and action constitutes only one aspect of a multi-faceted human activity, which nevertheless, thanks to its successes, tries to impose itself upon all other human activities. This philosophy demands the establishing of clearly marked limits and restrictions for technology.

Further away from the battlefields of these more specialised philosophies, the words *technophilia* and *technophobia* have emerged to characterise these two opposing views among 'ordinary' people. As discussed earlier, and still only roughly speaking, technophiles are mainly people directly involved in the development of natural sciences and technologies, and technophobes are mainly people coming from the humanities. Of course, there are always vested interests in taking one stance or the other, but, regardless of that, there is a real dilemma about the future of our techno-societies and it is certainly unwise just to ignore this dilemma. The topics of this discussion are very broad, even though they are often only related individually to some isolated aspects where a technological influence is judged as beneficial or damaging. We are obliged to discuss these topics in such general terms, however, because they underlie the whole development that is proposed and exemplified in the present work, and they may be expected to influence other works that may succeed the present one. If we then insist on discussing these matters in general terms, we see at once that there are several equally general questions that are nearly always raised in discussions about technological determinism. First, the question of the so-called *reification* of the word *technology* is often disputed. To 'reify' is to engage in objectification: to cast something into the form of an object, as a 'thing' that is taken up as a subject of predication. Even though we often use the word technology as a reference to one 'thing', one body, which comprises the totality of technological thought and action, it may be argued that this is not justified. Objections are usually raised by the technophiles, that there are so many different technologies, all with different purposes, addressing different goals, and even conditioned by different factors in their development. Without reification however, it would be extremely difficult to express our thoughts through any kinds of 'concepts'. Similar objections may be raised against many other terms, such as arise through the reification of the lexemes 'society', 'culture', 'politics', 'ideology' etc. The reification of 'technology', has similar origins and it underlines something common and unifying in all different technologies, regardless of their diversity, a something which allows us to put them all under the umbrella of one word.

The second question raised is the one of the *autonomy* of technology. This question is related to the widespread feeling that, even though technology is a human activity, in our times it has 'gained' a sort of autonomy, and its development has escaped human control. This stance is also widely disputed, again mainly from the technophilic camp, as unsubstantiated, and then not only because humans should always be able to control technological development, but also on the grounds that sometimes certain risks are necessary in order to achieve technological *progress*. The faith of the technophiles in technology is always expressed in their certainty that even if some 'unforeseen' consequences are spawned by a given technological development, there will always be some other, different, kind of technology which will resolve that problem. However, the feeling that technology has already an autonomy, and one which is far from being controllable by humans, spreads regardless of these assurances. This feeling is most appropriately described as a form of anxiety, rising up to the level of *fear*. It arises partly from our traditional thinking that the only locus for any autonomy is the human individual, and granting autonomy to any other agent leads to direct clashes with the conception of humans as individuals possessing the free will to make all relevant choices in their lives. Still, if the feeling persists, it is clear that there is something inherent in the development of modern technology that allows for this gradual increase in the autonomy of technology itself.

The third general point that is raised in relation to technological determinism is the question of the *neutrality* of technology. It is obviously linked to the previous question of the autonomy of technology. The proponents of technology's neutrality are again mainly recruited from the group of technophiles. They maintain that any technology, by itself, is neutral, and that the purpose for which that particular technology will be used only depends upon us. Neutrality of technology is usually coined in slogans like: 'technology only opens doors, it is up to us to go through these doors or not'; 'knives can be used to cook, kill or cure'; and 'guns don't kill, people do'. In contrast to this, the opposing opinion is that technology is far from neutral. In as much as we are responsible for shaping and influencing any technological tool, we are also shaped and influenced by the same tool. This opinion has especially gained weight with the deeper penetration into our every day lives of tools like computers, Internet, VCRs and television. Indeed, for many of us, our whole worldview has been changed through the introduction of these tools. The view of the non-neutrality of technology is coined in phrases like: 'to a man with a hammer, everything looks like a nail'; 'to a man with a camera, everything looks like an image'; or 'to a man with a computer, everything looks like data'. Even though the question of the neutrality of technology is still debated, this second view, of non-neutrality, and even the belligerence, of technology is nowadays the more widely accepted.

The whole point of technological determinism as seen *in action* is, however, that all these discussions will eventually come to matter very little. From this point of view, there is one crucial aspect of modern technology that actually shapes the meaning of the term technological determinism more than any of the other aspects mentioned above. It can probably best be expressed in the form of a *technological imperative*. In a sense it includes all the other aspects in it. It can be stated in the form of an argument that modern technology has already become so autonomous that it will finally itself defy any criteria for directing or limiting its development, *and precisely because there are no such criteria other than the criteria of technology itself.* In fact, although omnipresent in society, technological development is somehow also outside of society, and then exactly because it recasts *all* social problems in technical terms. By so 'technicising' all society it creates an infinite number of technical problems, but these in turn only require technical solutions.

The main criticism of technological determinism is aimed exactly at the technological imperative: the inability of any combination of technological means to resolve social problems that have themselves originated from the implementations of technology. The only result of the application of these means would be the creation of more problems, the solutions of which, in an endless and vicious circle (known as the 'technological fix'), would require the application of even more technology. Technology is something which only brings *new* social problems along with providing *progress*, but more often under the cover of providing the same kind of 'progress' it provides a reason more to promote its own interests. As Max Black (1980, in Bauman, 1996, p 186.) expresses this feature:

> I do not agree with the engineers and technologists who believe that the problems facing us are to be solved by the so-called technological fix... I happen to think that the problems raised by technological advance are probably insolvable.

Bauman (1996, p.186) explains the 'technological fix' to which Black refers as follows:

> In our age, technology has become a closed system; it posits the rest of the world as 'environment' - a source of food, raw matter for technical treatment, or the dumping ground for the (hopefully recyclable) waste of that treatment; and it defines its own misadventures or misdeeds as effects of its own insufficiency, and the resulting 'problems' as demands for more of itself; the more 'problems' technology spawns, the more technology is needed. Only technology can 'improve' on technology, curing yesterday's maladies with today's wonder-drugs, before their own side-effects set in tomorrow and call for new and improved drugs. This is probably the one problem 'raised by technological advance' which is fully and truly 'unsolvable': from the closed system: there is no exit. It is not so much the question of problems which call for more technology, but of the very presence of a technological capacity that cannot but 'problematise' aspects of the world which would not otherwise be seen as 'problems'. (that is, as 'wrong' states of affairs calling to be forcefully altered 'for the better').

The technological imperative constitutes technology's 'declaration of independence' of means from ends which necessarily leads towards a total sovereignty of means over ends. From then on it is no longer important what we aim to do, but only what we *can* do. It is the means that justify the ends. Just the availability of a new technology calls for its application and no other *legitimation* is needed for its application. To think that technology is there to satisfy needs and goals is then regarded as an illusion, and technology is seen as a self-propelling system which determines all other aspects of human organisation. This is similarly expressed by Ellul (1980, in Bauman, 1996, p.187):

> Technology never advances towards anything because it is pushed from behind. The technician does not know why he is working, and generally he does not much care. He works because he has instruments allowing him to perform a certain task, to succeed in a new operation...
>
> There is no call towards a goal; there is constraint by an engine placed in the back and not tolerating any halt of the machine...
>
> The interdependence of technological elements makes possible a very large number of 'solutions' for which there are no problems...
>
> Given that we can fly on the moon, what can we do on it and with it?...
> When technicians came to a certain degree of technicity in radio, fuels, metals, electronics, cybernetics etc, all these things combined and made it obvious that we could fly into the cosmos, etc. It was done because it could be done. That is all.

Ellul was probably one of the most profound critics of technological determinism. He identified it as one of the greatest threats to humanity. In his most important work *Technological Society* (1964) he warns:

> Each of us, in his own life, must seek ways of resisting and transcending technological determinism....The first act of freedom is to become aware of the necessity.

His great contribution was the identification of the one and only criterion that drives any development of modern technology: the criterion of *efficiency*. This is the quest for 'maximum yield' with 'minimum effort', in whatever way these happen to be defined. With this governing criterion, even the possibility for *real* diversity of technical means disappears: "the multiplicity of means is reduced to one: the most efficient" (ibid). More importantly, Ellul showed how this criterion has been applied to virtually all human activities in the process of their 'technicising', and how in this process these activities are being diminished. Efficiency becomes the governing criterion, and then not only for engineering technology, but also for scientific research, education, political activity, environmental protection or even cultural development. Humanity has sacrificed the *proximity* of all our relations, either among ourselves, or with our natural environment, to the criterion of efficiency, which results in all these relations becoming mediated through technology (ibid):

> ...the human race is beginning confusedly to understand at last that it is living in a new and unfamiliar universe. The new order was meant to be a buffer between man and nature. Unfortunately, it has evolved autonomously in such a way that man has lost all contact with his natural framework and has to do only with the organised technical intermediary which sustains relations both with the world of life and with the world of brute matter. Enclosed within his artificial creation, man finds that there

> is "no exit"; that he cannot pierce the shell of technology to find again the ancient milieu to which he was adapted for hundreds of thousands of years.
>
> The new milieu has its own specific laws which are not the laws of organic or inorganic matter. Man is still ignorant of these laws. It nevertheless begins to appear with crushing finality that a new necessity is taking over from the old. It is easy to boast victory over ancient oppression, but what if victory has been gained at the price of an even greater subjection to the forces of the artificial necessity of the technological society which has come to dominate our lives.

In this process, in order to satisfy the requirements of the technological imperative, it becomes 'necessary' to induce certain changes into the nature of humans themselves. One of these major changes is the transformation of our societies into 'mass societies', with needs and wants which are made so uniform that the existing technological products can satisfy them. The crucial role in this process of 'creation of needs and wants' is taken up by advertising enterprises. And, with that, the technologies of communication and the media, the so-called 'technologies of persuasion', gain a supreme importance as compared to any other technologies, as they became a sort of a 'fuel' for the 'technological engine'. No wonder that this process is often seen as an inevitable march towards some sort of technological totalitarianism (Norris, 1993).

Regardless of the accusation of taking an over-pessimistic view on technology, and for disregarding many apparent technological benefits, these views remain extremely relevant even in our present times. The crucial point, and our immediate purpose here, is however something very different. If we try and think about the criterion of efficiency, which indeed seems like the one and only criterion that governs any technological development, we shall discover that there is *nothing essentially technical* about this criterion. The origin of such a dominance of the efficiency criterion cannot be found in technology-as-such. In fact, in all of these criticisms of technological determinism, and even in the prophetic voices of empending 'doom', about the enslavement of humanity in a society governed by technology, the actual accusations cannot really be directed to technology-as-such, but only to what is widely called '*modern*' technology. 'Modern' technology is in this case regarded as something completely different from all other possible technologies.

In order to grasp the full importance of the distinction between technology-as-such and 'modern' technology, we will have to spend some time on that crucial word 'modern'. As will be argued here, that word carries much more of the 'doom and gloom' associated with the term 'modern technology' as compared to the second, 'technology'. The search for the origin of our present condition, which has required the total submission of all our thoughts and actions to the one criterion of *efficiency* while depriving all other possible criteria of their power to legitimise human thought and action, has then to begin with the unveiling of the *ideology of modernity*.

It is now time to recapitulate for the purpose of subsequent reference:

Thesis 2: *The constructs towards which we are striving must be directed to overcome powerful technocratic tendencies, both within society and within our own selves. These constructs should promote a healthy scepticism concerning the all inclusive claims of modern science and technology and demonstrate to us the inherent limitations of all claims to 'efficiency'.*

2.3 Modernity means separation

It can hardly be expected that any single characterisation of the project of modernity will receive a common consent. Since its very beginning (and it is even difficult to answer the question of *when* this beginning occurred: see Abbott, 1999b) the modern project has been judged in so many different ways that any attempt to 'present the whole picture' is impossible. There are several reasons for this. The most obvious one is the fact that within the modern era the world has experienced such radical changes, and has been transformed in so many different ways, that the sheer 'size of the subject' makes any attempt to provide an all-inclusive analysis impossible. Another reason, which is probably even more significant, is the very character of modernity, especially as it will be presented here. It has the character of separating, dissecting and compartmentalising our world into just so many different domains. In turn, each of these domains offers so many different interpretations that the difficulty of having one single picture arises precisely from this *modern* worldview, which is itself partitioned into segments of knowledge and segments of ignorance. The way in which modernity will be characterised here probably suffers from similar drawbacks. All that one can hope to achieve is to shed sufficient light on some crucial aspects of the modern project that it will be possible to rediscover sociotechnology.

The adjective 'modern' (*moderna*) has been continually used throughout history to mark any period abundant with some consciousness that in relation to the past it is a transition from the old to the new, usually presented as being from 'worse' to 'better', and so constituting 'progress'. Up until the period of the Enlightenment, modernity was always marked with some need for a renewed relationship with the ancients and the need to revive the antique past through some sort of imitation. This was a modernity that somehow always looked back for its inspiration. However, we shall concentrate here upon the specific part of the 'modern era' which accompanied the Enlightenment of the eighteenth century, which was different from previous periods in the sense that it gradually dissolved this relation to the classics of ancient times. With the exception of a romantic period at the beginning of the nineteenth century, with its turning back to an idealised Middle Age as its inspiration for a new historical era (modernity which again looks for inspiration to the past), the last three hundred years of human history have been marked by a modernity that is looking always ahead, growing into a position of an increasingly radical opposition between 'the traditional' and 'the desirable'.

The one word which is most commonly used to characterise this modern era is that of *separation*. In order to understand what we actually mean when we use this word as a distinguishing mark for this whole period, we will have to start from a central notion of the Enlightenment as a continuation and the fulfilment of the Renaissance: the replacement of the position of God by that of man[1] as the central element in the universe. A new order was to be constructed based entirely on human capabilities and suited to all human needs. However, these human 'needs' suddenly become something entirely different for the 'new', 'free' man. It appeared that by rejecting religious dogma and 'superstition' the possibility was opened for 'the pursuit of happiness' in this world, and the earthly life suddenly became a life of opportunity, instead of a torturous road towards salvation. Human desires, which were often suppressed as lusts and sins and, as such, were seen as violations of religious taboos, suddenly became potentially fulfillable during a regular human lifetime.

But, how was this new social order to be constructed? Once the Divine frame of guidance for thought and action began to dissolve, from where was the new source of authority going to proceed that would assure the achievement of a new and *stable* social order that offered each individual his own 'right' for 'the pursuit of happiness'? The new order was going to be a human order and this new authority was to be human *rationality*: 'reason as the almighty'. And the only way in which reason could proclaim this authority was through its access to knowledge, but then necessarily unhindered by any obstacles. "No doubt, the sovereignty of man lieth hid in knowledge", said Francis Bacon, in the early dawn of the Enlightenment (see Horkheimer and Adorno, 1944//1969//1973, p3). But this knowledge which was now revealed through human rationality became a different kind of knowledge as compared to any other knowledge. The new goal of establishing a human paradise in this world could only be achieved by transforming that same world (or in the beginning, at least, its inanimate part) into a pool of *resources*. Casting everything as some sort of resource required a different kind of knowledge. To 'know', gained a new meaning: to 'master' and to 'dominate'. This reason-mediated knowledge became *instrumental*. The world, emptied of any Divine presence, was at once at the disposal of the new rational man: first to explore and discover it and later to exploit it. Knowledge was to become the ultimate power. Any knowledge which could not provide this kind of contribution was to be immediately dismissed, simply because it did not empower its knower.

> Knowledge that tendeth but to satisfaction, is but as a courtesan, which is for pleasure, and not for fruit or generation. Not satisfaction which men call 'truth', but 'operation', 'to do business' is the right mark for what is the true end, scope or office of knowledge. (ibid, p.5).

---

[1] The word 'man' will be used continuously here to indicate what is supposed to be 'human', or even 'mankind'. We ask the readers not to consider this as some sort of gender bias, because it is done deliberately. Women were almost totally excluded from the 'Grand' project of modernisation, and were for the most part allowed to play only a 'supportive' role. And this 'support' was not usually directed to the project itself, but to the men who appropriated this complete project for themselves. Moreover, as Foucault in particular has emphasised, the concept of 'man' is itself a product of 'modernity'. In fact, this effective exclusion of women from participation in the enlightenment is of the greatest significance. Indeed, we can ascertain that this programme would have been impossible of accomplishment without this exclusion. Several decades of study of 'Gender Issues' has now made this abundantly clear, while the reasons behind it (its 'logic') have become quite evident (e.g. Abbott, 2000a).

The new, 'free' man who was about to bring the ideal social order into being could only do this by separating himself from a non-human nature in order to dominate it. The two tasks - creation of a new social order and control and dominion over nature - required this clear separation between *society* and *nature*. Nature was the part of the world that was 'out there', non-human (although plentiful of life), and it was completely at man's disposal. On the other hand, society was to become a pure creation of this 'free' man and in the realm of society there was no place for nature, be it inanimate or not.

In spite of the paradoxes that soon emerged from this separation, it was necessary to keep these two separated in order to keep faith with the modern ideology. The first paradox resulted immediately from the development of the modern sciences (especially modern physical science), which found a fertile ground for its development in this separation. Namely, although nature was to be seen as transcendent and separated from the human, by virtue of the scientific experiment carried out in the laboratory it become possible to fabricate 'facts' about nature, 'facts which spoke for themselves' although it was the scientists who translated the speaking. Suddenly nature was not only 'out there' waiting for us to discover it, but, since it persisted in surpassing us infinitely, we could never complete our efforts to reconstruct it in thought. On the other hand, a similar paradox occurred at the other pole, of society. Although society still appeared to be constructed entirely by humans, it became increasingly sustained through the employment of an endless number of objects, things, non-human entities, that themselves increasingly took the forms of artefacts. It was through this process that society became transcendent, surpassing man alone, despite his appearing as the one-and-only actor in making that society.

These paradoxes are at the centre of what Latour (1993) has called the modern 'constitution'. This 'constitution', as any constitution should do, provided guaranties, and in this case, that, in spite of its inherent paradoxes, a proper separation would be maintained between nature and society. Although, in fact, we are today dealing with a great number of 'hybrids' (humans in relation with human, non human natural entities and human made artefacts etc.) which are situated between the two, now widely separated, poles, the existence of these 'hybrids' is never to be admitted. Latour (1993, p.32) explained what this ingenious construction of the modern 'constitution' provided:

> ....(The moderns) are able to make Nature intervene at every point in the fabrication of their societies while they go right on attributing to Nature its radical transcendence. They are going to be able to become the only actors in their political destiny, while they go right on making their society hold together by mobilising Nature...
>
> We must admit that this is rather neat construction that makes it possible to do everything without being limited by anything. It is not surprising that this Constitution should have made it possible, as people used to say, to 'liberate productive forces'...

By maintaining the belief that there is a proper separation between humans and non-humans (human society and nature) and at the same time cancelling out this separation, the modern idea became apparently invincible. The eternal questions of truth, justice and beauty were taken over from the supreme Divine judgement and placed in the hands of man, and the search for the universal answers to these questions became the tasks of three separate spheres: science, ethics and art. Within these spheres, separate 'governments' were created, each one responsible for this search for 'universal' answers. Although the practices of these 'governments' in their quest for "objective science, universal morality and law and autonomous art according to their inner logic" (Habermas, 1985) were constantly realised through the introduction of the 'hybrids', this was never to be accepted, and officially these spheres were maintained separately.

One consequence of this separation was the provision of a fertile ground for the development of modern sciences. Indeed, it was in accordance with this separation that the separate sciences emerged. Each particular aspect of the world was henceforth to be covered by a scientific field. Even within one branch of, for example, physical science, the overlaps were rare, and any such overlaps with the social sciences were unimaginable. The University, contrary to what the name suggests (*universal*), became a site for institutionalising this departmentalisation of scientific knowledge.

Regardless of the emergence of new and different scientific fields, these were all supposed to be united in their following one and same *scientific method.* A method which, although essentially simple, proved to be extremely successful in discovering 'scientific truths'[2]. To discover 'scientific truth' within the modern constitution meant that by applying logical reasoning within a given axiomatic system (the essence of the scientific method), supported by a systematic gathering of evidence through scientific experiments, man would be able to unveil the hidden *laws* that govern nature. The isolated, 'cold' and 'foreign' 'nature' would in this way become accessible to man because its laws could be mastered in a form of scientific *theories*.

The spectacular development of science that followed, brought a new kind of separation. Scientific knowledge became the only kind of knowledge which could deal with the determination and application of the criterion of truth, or more precisely 'scientific truth'. The procedures of standard logical reasoning which were seen as underlying all research (production) and transmission of scientific knowledge, apparently became the only ways in which any truth could be discovered. All other knowledge, which was not amenable to the processes of argumentation and proof, was to be immediately discarded. Even in our common vocabulary, the word *knowledge* was, until very recently, associated exclusively in its meaning with *scientific knowledge*.

In our investigation of the ways in which this supreme position of scientific knowledge over any other knowledge works, it is important to notice immediately that knowledge is more than science and even more than learning (Lyotard 1979//1984). Each individual and any community possesses a body of a different kind of knowledge, which is sometimes called 'traditional' knowledge, or 'custom' knowledge or even, more restrictively, 'common sense' knowledge. When Lyotard (1979//1984) investigated the nature of this knowledge and its

---

[2] 'Scientific truth' is essentially truth defined as 'correspondence', usually between two 'statements'. Abbott, (1991, p.4-5) points out that this view of truth as correspondence has its origin in the scholastics, and is derived directly from Christian philosophy.

relation to scientific knowledge, he used the term *narrative knowledge* to describe it. This term was particularly well suited for his purpose because in his investigation he studied knowledge in general in terms of discourse, so that the relations between different kinds of knowledge proceeded through identifying the pragmatics of the associated language games. But indeed many authors have agreed that the narrative form is pre-emptive in the formulation and transmission of this 'traditional' or 'custom' knowledge.

This 'narrative' knowledge deals with the establishment and application of other criteria then scientific truth, such as other criteria of justice, other criteria of beauty or even other criteria of efficiency, even though the last has been taken over and cast into a scientific form so as to become the most important set of criteria in societies faithful to the modern ideology. In these processes it is the narrative form itself which is used or exploited, both for constituting and implementing the 'appropriate' criteria. There is however no 'scientific method' which can itself be learned and then applied for this kind of purpose: in this way the *legitimation* of narrative knowledge comes only from its conformation with the accepted criteria in a given social circle, a consensus of 'opinions' which are themselves constructed through the use of the narrative form itself.

Modern ideology separates scientific knowledge from narrative knowledge and establishes one specific and highly unequal relation between the two. Scientific knowledge very often turns to this 'narrative' knowledge and tries to apply its methods of argumentation, validation and proof, but then discovers more often than not that 'narrative' knowledge is not subject to these methods. Science then condemns it as knowledge which is unscientific (which in the modern sense of 'scientific' is true), but then, because of that, in some way 'inferior'. "Therefore it classifies it as savage, primitive, underdeveloped, alienated, composed of opinions, customs, authority, prejudice, ideology, ignorance etc. Narratives are fables, legends, myths, fit only for women and children etc." (Lyotard 1979//1984, p.27). Because of this perception of 'narrative' knowledge, science sees itself directed to 'enlighten', 'civilise', 'educate' and 'develop'. However, the accomplishment of this programme then calls for regulating, prescribing and legislating from the side of modern science.

This then brings us to another kind of separation brought about by modernity: the separation of people. Accepting the idea that society is to be build entirely by humans and on the basis of their human nature, the philosophers of the Enlightenment were immediately puzzled by one contradiction. On one side the new order was to be based on the 'nature of man', but when they turned towards 'empirical' men and women they became fearful, it being completely unacceptable to them that this degraded nature, as it stood, could provide such a source. This was only partly because, by accepting it, their spiritual leadership would be undermined, and their services made redundant. (which indeed they became in certain more violent periods, such as during the latter part of the French revolution!)

The solution was found in proclaiming that human nature was indeed the foundation, but that this was not the nature which was present at that moment, but an idealised one which existed only potentially, one that was 'not yet': an unfulfilled potential that was ever-unfulfillable of its own volition but that might be realised with the assistance of reason and so through the intercession of the reason-bearers (Bauman 1996). The 'elite', the 'knowers' were thereby immediately emancipated from 'the masses', quite separated from these bearers of the 'animal' and 'original-natural' in man, these dependent humans who were only guided by their impulses. At the beginning this emancipation was only a 'closing-of', a distancing accompanied even by a loathing, but it was followed by the perception of the 'others' as 'objects of rule and care'. These two tasks, when combined, created the possibility of 'political leadership'. Immediately the separation of the rule-makers and the rule-followers seemed inevitable and necessary. These philosophical ideas were taken over by legislators who could then combine coercive power with persuasion by opening the eyes of the populous to the advantages of following the rules 'for their own sakes', and these persuasive elements became the building blocks of the new social order. As soon as a glorifying 'free will' appeared as a possibility for salvation, it promptly vanished into a 'legislative assembly'. The new human freedom turned into the freedom of the chosen few. Bauman (1996, p.27-28) explains this 'curse of modernity' as follows:

> Reason is a shared property, but in the case of this particular equality, as in all other cases, some humans are more equal than others. Philosophers are the people endowed with more direct access to reason, the unalloyed reason, reason unbeclouded by narrow interests; it is their task therefore to find out what sort of behaviour reason would dictate to the reasonable person. Having found this out, they should communicate their findings to those less endowed, who cannot find them on their own, and do it with the *authority* of the 'people in the know'. To the others, to whom the message is addressed, the findings come, however, in the form of Law: not a rule inherent in their own choices, but one that prompts the choice from outside. Despite the fact that reason is every person's property, the rules promulgated in the name of reason are to be obeyed after the pattern of submission to an overwhelming external force. They can be best thought of in the way we think of laws legislated by authorities armed with coercive means to enforce their instructions...
>
> People must be told what their true interests are; if they do not listen or appear to be hard of hearing, they must be forced to behave as their real interest demands - if necessary, against their will...

The rules were of course not universal and everywhere applicable, but each idea behind the derivation of these rules was striving exactly towards this: its universality. Although it has been repeatedly shown that there are failures and clear social deadlocks, these have always been treated as 'temporary' and only 'necessary for the purpose', perhaps of learning some lessons, so that the march towards the new universal human order had to continue. What was universal on the other hand for all of these 'Grand Narratives' was the disqualification and exclusion of those who would not follow or obey their rules. The grand idea of putting man in the centre of the universe through a separation of nature and society brought with it one of the greatest separations between humans in history.

If we now go back to the discussion of the separation of scientific knowledge from any other kind of knowledge in relation to the above presented separation of people, we can see how the separation of scientific knowledge from narrative knowledge is just as paradoxical as is the separation of nature and society. On the one hand, in modern societies, science has everything to do with establishing what is true but has little or nothing to do with establishing what is just. On the other hand, in modern society, rules of conduct within society and the search for justice and morality henceforth have nothing to do with establishing scientific truth. Despite of all this, however, science is allowed to intervene at any time and place and make rules for the regulation of society, simply because science becomes the pure epitome of rationality, that supreme authority which the new man holds in his own capacity. Therefore, rules which cannot be immediately rationally justified lose any applicability, and people who refuse to follow rules legitimised by rational–scientific knowledge can only have one choice: accept or be marginalised. The instrumental nature of rational-scientific knowledge with its origin in the quest for mastering and dominating nature becomes extended to the task of mastering and dominating people as well.

It would certainly be wrong to think that this whole development in the modern era was intentionally master-minded by isolated malicious groups and individuals guided exclusively by their lust for power. The Enlightenment, identified with the philosophy of rational humanism which brought along with it the dominant role of scientific knowledge, certainly did not come 'out of nowhere', or as a result of human opportunism. It was the culmination of a long process, and indeed one that lasted for centuries, with its origins stretching back at least to the early dawn of the Renaissance. The painstaking development of modern physical science, first through its geometrical period, inevitably followed by its symbolic (algebraic) period culminated in the seventeenth century with the master works of Galileo, Descartes, and Newton (for a most comprehensive summary of these developments see the works of Duhem ,1913-1959//e.g. 1985; but also Abbott, 1999b). By that time, the art of combining systematic experimentation with deductive reasoning from a small number of basic principles was establishing itself as the only method with the potential for discovering the laws of nature.

However, the application of this methodology did not remain restricted to the domain of natural science. Almost simultaneously in that same century, two great thinkers, Spinoza and Hobbes created their own masterpieces, in ethics and political philosophy respectively. And both of them were inspired by the *same* rational-scientific method. Spinoza's *Ethics* (1677//1983) is constructed entirely along the lines of the Euclidean 'geometrical method' of argumentation and proof, this being for the author the only appropriate method of objective rational enquiry. Hobbes' *Leviathan* (1651//1961) does not have the same strict logical form, but, again, his whole system of political philosophy is constructed by a careful and structured deduction from a small set of first principles.

But it is not only the *method* that characterises these two works as truly modern. Spinoza's pantheistic worldview identified God with nature and, by that, God became immanent to man, instead of being some external 'cause of all things', transcending both man and the world. And Reason (and now with an upper case 'R') was the only human capacity, which would allow not just for the cognition of 'truths' of nature (and with that of God), but also for cognition of ultimate 'good'. Rationality was proposed not just as the proper method for thought, but also as the proper way of life. Only rational faculties could lead to human happiness and freedom. What is similar in both Spinoza and Hobbes is the identification of the urge for self-preservation as the essence of human existence. But, while Spinoza finds that this will be best achieved if man follows the orders of Reason *because that means reaching man's true 'natural' state*, Hobbes finds that Reason is the only *salvation from man's 'natural' state*, which is of chaos, brutality and violent death, resulting from his uncontrolled desires and passions. Here Reason must be applied to create an 'order', which is a social order. Hobbes empties human society from any Divine presence and Reason must be applied to create the new 'institutions of common-wealth', where human desires would be translated into rights and duties, and thereby controlled. This difference becomes understandable when considering the different projects that these two thinkers were set to conduct. Spinoza creates a complete philosophical and ethical system *based* on rationality, and Hobbes creates a new social system *maintained* by rationality.

In both cases, their ideas meant a radical departure from the established orders in their own societies. As a result, their works were condemned by many of their contemporaries as atheistic, blasphemous and profane. But, at the same time, they opened the way for Reason and human rationality to extend their object of enquiry from the inanimate world of nature into the world of social relations among humans. The stage was being set for the *ideology* of the Enlightenment: the sincere belief that humanity, by applying reason, is on its way to emancipation, and that it thereby starts progressing towards the ultimate ideal of a perfect social order.

We have identified several separations that mark the modern era, these being those which are most immediately related to our main task in this chapter, of rediscovering sociotechnology and our task in this work generally of identifying and exemplifying the means to overcome these separations. Separation of man from God, separation of society from nature, separation of scientific knowledge from any other knowledge and separation of people: above all these separations stands Reason, now indeed promoted as something 'almighty'. And, at the same time, as Latour indicated, they are all just as powerful as they are also paradoxical. The transcendent God has been removed from the modern world consisting of nature and society, but He can still descend into men's hearts. He cannot intervene in their external affairs except through their own volition, but they themselves can still remain spiritual. Nature is separated from society, but the new society is itself only made durable by mobilising the forces of nature. Scientific knowledge becomes a supreme knowledge, even though this supreme position is ideologically guaranteed only within the Grand Narrative of modernity. Modernity brings a new and ideal order to the whole of humanity, but some are responsible for carrying out this task, while others are obliged to follow. It seems that the power of modernity *resides* in these paradoxes, or as Latour (1993, p.33-39) puts it:

> You think that thunder is a divinity? The modern critique will show that it is generated by mere physical mechanisms that have no influence over the progress of human affairs. You are stuck in a traditional economy? The modern critique will show you that physical mechanisms can upset the progress of human affairs by mobilising huge productive forces. You think that the spirits of the ancestors hold you forever hostage to their laws? The modern critique will show you that you are hostage to yourselves and that the spiritual world is your own human - too human - construction. You think that you can do everything and develop your societies as you see fit? The modern critique will show you that the iron laws of society and economics are much more inflexible than those of your ancestors. You are indignant that the world is being mechanised? The modern critique will tell you about the creator God to whom everything belongs and who gave man everything.

You are indignant that society is secular? The modern critique will show you that spirituality is thereby liberated, and that a wholly spiritual religion is far superior. You call yourself religious? The modern critique will have a hearty laugh at your expense........

Never has a Constitution allowed such a margin for manoeuvre in practice."

In effect, the modern 'constitution' requires that one and the same person holds self-contradictory beliefs at one and the same time, which corresponds to a state of *aporia*.

Two instruments have shown to be the most enduring in their suitability for introducing the rules of separation and securing their implementation, while interchangeably utilising and managing such a wide 'margin for manoeuvre': the *democratic nation-state* and the *market*. They are both still very much present. There have been significant internal changes and their appearance is nowadays different from that of the period of their establishment, but they are still the two basic instrumental institutions around which all crucial aspects of our modern condition are centred (Dunn, 2000). They have been working through history, both in union and in opposition, but always in some sort of conjunction. However, it now more often appears that the market is going to take over, as the processes of globalisation gradually dissolve the economic, cultural and military foundations of the nation states. (This process of globalisation is only marginal to the purposes of our present immediate discussion, but we shall return to it later in a somewhat different context).

At the end of this section we should return to the question of the 'technological imperative', or the 'technological fix'. First of all, so far in our discussion we have largely ignored the crucial role which technology played in the development of the modern project. Most of the developments throughout the so-called geometrical period of the modern era were technologically driven, and there was no support from any kind of 'natural science' (see again Abbott 1999b). There are numerous examples of 'pronouncements after the facts' throughout the centuries, which indicate how technological developments preceded their explanations and especially their explanations in terms of scientific theories. Up until the period of Enlightenment, technology existed as an integral part of the human lifeworld and, as with any other domain of life it was not formalised in a scientific manner, whether as a concept or as a social force. But, the processes of separation that came with modernity inevitably needed to employ technology in a different manner. This difference in manner meant that technology had to be used for modern 'purposes' and the view was gradually formed that technology was 'applied science'. While science kept for itself the descriptive role, concerned 'solely with how things are', technology was transformed into a *vehicle* of the 'normative', concerning itself with human goals and questions of how things 'ought' to be. All this proceeded regardless of the fact that it has been shown, at least since Kant and as repeated by so many others after him, that there is no straightforward *logical* transition from an 'is' to an 'ought'. From scientific denotative statements (facts) which may be true or false, there is no direct route to prescriptive (imperative) statements which guide human action. Technology as 'applied science' nevertheless acquired this capacity, this role, and the absence of any logical transition was substituted by the modern ideology of *scientific progress*.

We shall concentrate more on the essence of technology in the last section of this chapter, but it will suffice here to say that the view of technology as 'applied science', or 'applied modern science' is largely misleading. The now common distinction between premodern, modern, and even postmodern technology, has nothing to do with the essence of technology. Technology as such, therefore has nothing to do with the 'doom and gloom' perspectives expressed earlier in the form of the 'technological fix' or the 'technological imperative'. We might even better use the terms 'state-fix', and nowadays increasingly 'market-fix' and 'market imperative' (or simply 'modern imperative'). It is not technology which is 'pushing from behind' towards more technology, but the modern rules of separation that have created and nourished this movement through the institutions of the nation-state and the market.

It is also important to note that we can easily extend the list of kinds of separations with a number of others that are closely related to the 'modern condition'. Man / women; physical / mental; spiritual / material; civilised / savage etc. Some of them are also related to this whole study, but the ones identified above are seen as the most immediately relevant. Specifically, the separation of knowledge and the separation of people will be of our main concern when dealing with development of new kinds of technical means for improving the processes of decision making in the field of water and environment. But, to be quite clear about this, we are not stating here that before the 'modern era' there were no such separations at all. As Derrida has shown (see, in this respect Powel 1997) throughout history human thought has always been based on some kind of separation, with one concept or entity placed in the centre while at the same time defining its central position through a marginalising or excluding of the other(s). What came with modernity was the *institutionalisation* of these separations, yielding both positive and negative forces unprecedented in history.

Let us paraphrase this into a further Thesis for later reference:

Thesis 3: *We have to aim for technical devices and social arrangements integrated in such a way as to expose and overcome the separations that the modern 'constitution' has introduced into our thinking. In particular, the separation between narrative knowledge and (modern) scientific knowledge has to be overcome: our new devices and institutions will have to find means not only to accommodate but to amalgamate these forms of knowledge to provide synergies which can treat the aporiaic tendencies in society.*

2.4 Return of the hybrids

There is no doubt that the modern era with its ingeniously (self-)constructed 'constitution', augmented the power of the 'Western' societies to an extent which enabled them to dominate both the world of nature and all other human societies which did not follow the same modern ideology. The history of this double act of colonisation has been recounted any number of times and is nowadays only too well known. These so-called 'Western' societies (even though they came to include Japan!) were, however, at the same time radically transformed as a result of the clashes between the ideology of Enlightenment and the existing ethical values firmly established during earlier periods. From that particular perspective, the emergence of the new concepts that followed from these clashes are commonly regarded as truly liberating. Let us try to present a selected list of these concepts by borrowing from Galagher (1998):

> -people came to be seen as having inalienable or built-in "rights", against which [the governing] authorities could not assert their prerogatives, and remaking society to protect those rights seemed like a logical next step.
> -blind obedience ceased to be a virtue, and instead came to be seen as an undesirable sign of slavishness
> -scientific investigation of the world was a good thing to do, since the material world was no longer seen as the providence of the devil
> -sexual desire became a good thing to have, especially after Freud. No longer was sexuality seen in the context of sin and [a] disruption of [the] social order, but was linked with mental health and creative energy
> -originality, creativity, individual brain-power were seen as powerful and effective, no longer overshadowed by the creativity at the beginning of time accomplished by God *ex nihilo*
> -self esteem came to be considered a positive, [and was] no longer labelled [as] the sin of pride
> -the world's various religions and philosophies came to be studied anthropologically, which put them all on a par. Each was "true in its own way", paving the way for ethical and theological relativism. Atheism came to be seen as just another possible answer to the riddle of the universe.
>
> Materialism, anti-authoritarianism, human rights, the idea of the individual, "the emergence of the body". All of these burst on the scene almost all of a piece. The Great Chain of Being would never be the same again.

Most of the concepts listed above stand as foundations of the value-systems in all societies that followed the modern project, or were later 'modernised'. In the light of these concepts it is very difficult for most persons to see what is wrong with this great project, or even to suggest what a better alternative could be. Although a line of critical thought has continuously followed the development of modernity, the above concepts cannot be easily challenged in our present-day societies.

Yet, at the same time, we are living in a condition where maintaining human rights still demands a continuous and apparently endless struggle; the authority of the state or the market demands new kinds of obedience, science-driven developments are blamed for unprecedented natural destruction, and individual originality and creativity is too often subjected to the power of the corporation. It seems as though many modern concepts have turned out to be only ideals, if not actually illusions. Humanity somehow finds itself trapped in the workings of a system that was constructed in the name of human liberation but often seems to be anything but liberating. In fact we have reached a stage of modern development where the inherent contradictions and paradoxes of modern society cease to play their sustaining role for the project. Instead they become destructive, or at least become more openly disclosed, demanding serious rethinking. The present condition is characterised exactly by this situation, and some truly exceptional events are now forcing this kind of re-assessment.

Our society is most often described as a postindustrial society and our contemporary culture as a postmodern culture. Very roughly this marks the realisation that the project of industrialisation is limited and that at least we have doubts about the success of the project of 'modernisation'. Historically it coincides with two important events: the end of the Cold War with the definite fall of 'socialism' and its elimination as a threat or alternative to Western capitalism and its liberal democracy. At the same time, the so-called 'West' is facing the fact that the scientific and technological developments nourished in its own institutions, through the rules of separation, have become so transcendent that now they raise questions about the global state of the planet and our own survival. In a certain sense these processes mark the end of the Enlightenment and industrial-social (and now essentially capitalist) hopes for unlimited conquest and dominion over nature. Victory and defeat at the same time! Latour (1993, p.9) describes how these conditions lead towards what he calls a 'double tragedy':

> The perfect symmetry between the dismantling of the wall of shame [The Berlin wall] and the end of limitless Nature is invisible only to the rich Western democracies. The various manifestations of socialism destroyed both their peoples and their ecosystems, whereas the powers of the North and the West have been able to save their peoples and some of their countrysides by destroying the rest of the world and reducing its peoples to abject poverty. Hence a double tragedy: the former socialist societies think they can solve their problems by imitating the West; the West thinks it has escaped both problems and believes it has lessons for others even as it leaves the Earth and its people to die. The West thinks it is the sole possessor of the clever trick that will allow it to keep on winning indefinitely, whereas it has perhaps already lost everything.

This situation has raised the doubt: Where does 'modernisation' lead? Where does a rationally-governed separation lead? The explosive development of the sciences and the technologies which was made possible through the modern separations has reached the point where it unveils the illusive nature of these same separations. Scientific knowledge is anything but separated from 'other' knowledge. When reports came out that advances in genetic engineering had first led to the successful cloning of a sheep, and then to the initial successes of a complete decoding of the human genome, a world-wide network of state-political and market institutions was shaken. The reactions have ranged from appeals for the banning of further experiments on ethical grounds and the mobilising of all kinds of 'other' knowledge to insure the implementation of this ban, to insurance companies seeing unforeseen possibilities for them to increase their profits with less risk if they can have access to the genetic code of every individual. This is a science that leaves nobody untouched and uninfluenced: employers, hospitals, military officials and planners, ordinary people. Global warming, ozone layers, deforestation (let alone the monstrous achievements of the weapon industries) and similar issues are all translated through similar 'networks'.

The 'hybrids', who's existence was never to be admitted by the modern constitution, have returned, but now as monsters. In fact they have always been here; it is only that the stakes have become so big that they cannot remain hidden any more. Our present world is constituted by these sociotechnological networks ('hybrids') and we have to realise that we are reaching the point where we cannot approach them separately anymore (even though this was the guarantee of 'progress' under the modern constitution). Knowledge, power, justice and interest are already mixed in these sociotechnological networks and we have to approach them as such.

This new condition has slowly begun to influence the ways in which we understand and employ scientific knowledge, both in its production and in its transmission. The boundaries between the scientific disciplines are gradually dissolving and new fields are emerging between them. The phenomenon of 'interdisciplinary studies' has evolved from a merely fashionable one into a real and necessary practice. The universities gradually lose their traditional role in our society as institutions through which scientific knowledge establishes its legitimacy. Their faculties dissolve into all kinds of new institutes and foundations where they merge to create new fields. In particular, the production of scientific knowledge through research increasingly leaves the universities altogether and in many cases these remain as institutions which guarantee only a production of teachers and not of researchers (Lyotard 1979//1984).

It is commonly observed that this space 'in between' the traditional scientific disciplines is more densely covered with new fields in the area of the natural sciences, but in fact much the same situation holds for the social sciences as well. A number of sociologists and social psychologists have recently recognised the need to include human artefacts in the scope of their sciences. It has been increasingly shown that these 'objects' have in fact a profound influence on the whole social process. The range of these artefacts when seen from this perspective can vary from "portable and wearable objects like books, clothes and jewellery, larger things like computers, cars and trains, settings and environmental features such as rooms, doorways, signs and advertising hoardings, and systems and processes like photography, the telephone systems and electronic mail" (Miller 1995). Although this list seems much too inclusive, it is not difficult to realise that indeed each of these kinds of objects has its place in the social process. Some of them require or allow or facilitate a certain social behaviour, others affect people's agency, while with others people indeed interact socially so as to provide them with agency, and even personality. It is increasingly understood that these 'objects' can no longer be discarded from the field of social sciences, as the modern constitution requires (Latour, 1993). Their traditional identification as 'objects with inner properties that do not count' in the social process, just 'screens' for projecting social needs and interests, shaped exclusively by relations among humans, is gradually abandoned, as the realisation spreads that society is in fact 'held together' by its artefacts (Abbott, 1996).

In parallel with this recognition of the social role of inanimate objects, and specifically human-made technological artefacts, the very practice of doing research in social sciences has in turn been changed by the new possibilities opened up by advanced technologies such as computer simulations. This development has reshaped such fields 'from inside'. Social sciences have always been faced with the insurmountable difficulty of not being able to carry out experiments in a 'social laboratory' in the way that the natural sciences were able to carry out their research. That the nature of this difficulty goes beyond the practical limitations of 'social surveys' is described by Lyotard, following his method of analysing scientific knowledge by identifying the pragmatics of language games (1979//1984, p. 57):

> It is generally accepted that nature is an indifferent, not deceptive, opponent, and it is upon this basis that the distinction is made between the natural and the human sciences. In pragmatic terms, this means that in the natural sciences "nature" is the referent - mute, but as predictable as a die thrown a great number of times - about which scientists exchange denotative utterances constituting moves that play against one another. In human sciences, on the other hand, the referent (man) is a participant in the game, one that speaks and develops a strategy (a mixed strategy, perhaps) to counter that of the scientist: here, the kind of chance with which the scientists is confronted is not object based or indifferent, but behavioural or strategic – in other words, agonistic.

Although this remains a serious problem (and at least potentially a problem of self-reference) social sciences have found new ways of dealing with it by using *simulations* of social processes using computers. In the beginning the approach taken was that of applying the traditional social theories, but very soon it proved possible to built artificial societies which become emergent from the complex interactions of simulated individuals (agents). Regardless of the crudeness of the representation of the individuals, this technology has opened new avenues for experimentation for many fields in the social sciences. Social simulation is actually becoming one of the most exciting and certainly the fastest growing sub-fields of sociology (see, for an introduction, Gilbert, 1996, and, more recently, Edmonds and Dautenhahn, 2001). Other fields, such as in law, art, psychology, and even philosophy are experiencing similar kinds of transformations, all pointing towards the *inseparability* of human-human relations from relations between humans and their artefacts.

In a similar way, the whole environmental movement can be seen as a rejection of the separation of humans and non-humans. Our fellow creatures- animals and plants as well as other non-living environmental features - are seen as integral parts of our new 'society'. New fields emerge related to this new perception. The field of Environmental Ethics (Armstrong and Botzler (1993)) is a particularly indicative example, where people try to mobilise traditional scientific knowledge along with all kinds of 'other' knowledge (moral philosophy, aesthetics, history, economy, politics, law, ecology, feminism and religion) under one framework in order to investigate this rediscovered union of ourselves with our natural environment[3].

The awareness that some boundaries must be crossed in order to see if they really exist, spreads even further than the separation between nature and society, and between scientific knowledge and 'other' knowledge. The disregard for the interconnectedness of humans, their tools and nature has led to environmental disasters of unprecedented scales. This situation has led to something unthinkable during the modern quest for progress: an open discussion between science and religion. From the very beginning of the environmental movement,

[3] Although the expression 'natural environment' is nowadays introduced in order to distinguish the 'natural' from so many 'artificial environments', it somehow signifies this whole awakening process that we are discussing. In the modern vocabulary the word 'nature' was replaced with the word 'environment'. This brought the feeling of loss of *proximity* with nature. The word 'natural' is still used to express a closeness that rejects any need for explanation, while 'environmental' is reserved for describing influences 'from outside' for which *relations* have to be identified and explained. 'Natural environment' tries to reconcile this difference.

reminders of various religious perspectives were offered as support for the environmental 'cause'. They were always received with respect, but rarely taken into much account in the 'big discussion', which always evolved around 'environment and development'. However, even these attitudes are beginning to change. A recent symposium entitled "Revelation and the Environment" has brought together about 200 scientists, religious leaders, philosophers, economists, artists and policy makers. This was described as "an important step in the long-running effort to find common ground among religious and scientific leaders who share similar concerns about the environment, but whose historical antagonism has often blocked collaboration" (Hobson and Lubchenco, 1997, p. viii). Some of the main conclusions reached by the 'scientific committee' of this symposium sound strikingly anti-modern (ibid):

> Despite incontestable evidence of the worsening state of our environment and its grim implications for the future, scientific discourse appears insufficient to produce personal or social action on a significantly large scale. To the extent that religion, with its millennial tradition of access to moral and symbolic dimensions of humans, can espouse the accompanying imperative of scientific observation, it can endow society with a new vision that can lift science from its isolation as a social force and promote it into a materially-based pillar of moral and spiritual existence.

Anybody truly faithful to the modern project, may be appalled by this 'surrender' of science. At the same time, a theologian who attended an earlier gathering of a similar kind ('Joint Appeal by Science and Religion on the Environment', 1992 hosted by US senator Al Gore), initiated by world-renowned scientists, expressed the unease occurring on the other side of the discussion table (Haught, in Daly, 1997):

> (I wonder) whether it is completely honest for them (the scientists) to drink in this case so lustily from the stream of moral fervour that flows from what they have consistently taken to be an inappropriate and even false consciousness of religious believers...the well-intended effort by the sceptics to co-opt the moral enthusiasm of the religious for the sake of ecology is especially puzzling, in view of the fact that it is only because believers take their religious symbols and ideas to be disclosive of the *truth* of reality that they are aroused to moral passion in the first place. If devotees thought that their religions were not representative of the way things *really* were, then the religions would be ethically impotent.

In spite of these examples which point out that this discussion will not be easy at all, both sides are beginning to recognise that there is hardly any other alternative for dealing with burning environmental problems. It will certainly mean that a lot of perspectives will have to change. And it is not a question of 'modernising the religions' and asking them to 'walk in step with contemporary reality', or 'suppressing scientific advancement to stop the continuous destruction of our planet'. Both sides will probably have to undergo some significant changes in order to reach some new, more reconciling paradigm, which may provide a crucial support for a much-desired sustainability.

What this new paradigm will bring, and even how it can be reached, is difficult to predict. The only thing that for the time being has been realised is that it is necessary. It should be stressed that this necessity cannot be viewed only within the context of contesting the modern world-view. While the limitations of the strict separation between 'a rational mind and a spiritual heart' are now disclosed, the new paradigm is also needed as the processes of collapsing the separations between humans, the natural, and the artificial develop. These processes can contribute little in dealing with our current social, environmental or technical challenges, and even further aggravate our condition, if they are controlled and governed only by apparently omnipotent, but essentially blind, market forces.

The new paradigm will have to reintroduce *spirituality* in a new way, which will have to be a way that does not oppose the realm of science. It can be foreseen that this process will face many difficulties. On one side, the very concept of spirituality is frequently rejected by traditional religions, as a poor substitute (created by the modern ideology) for true religious belief (see Nasr, 1999), and therefore rendered impotent. On the other hand, an even greater difficulty arises, which is for the paradigms of spirituality to be transformed into 'metanarratives', which are immediately rejected in our postmodern culture, which is even defined by Lyotard (1979//1984) as a culture with an inbuilt 'incredulity towards metanarratives'. One indicative example of this occurrence arises in the field of 'eco-feminism', which in the beginning adopted a triadic paradigm of feminism, ecology and spirituality framed in the "Goddess-nature-principle" and the "Gaia-metaphor" (see for example Spretnak, 1996, Shiva, 1996). Within this framework, nature was introduced both as 'alive' and 'sacred', and our mutual relationship with it as one of 'motherly love and care'. As soon as this framework created a powerful critique of the instrumental reason (identified with patriarchy by eco-feminists), and even became compelling for other social groups, in a way that could generate significant political action, it became self-deconstructing. The same symbolism and the resulting discourse which caused the success were challenged by eco-feminists themselves as in fact oppressing both women and nature (Warren, 1996):

> Language which so feminises nature and naturalises women describes, reflects, and perpetuates the domination and inferiorisation of both by failing to see the extent to which the twin dominations of women and nature (including animals) are, in fact, culturally (and not merely figuratively) analogous.

A similar view, and one that is even more directly related to this 'incredulity towards metanarratives' is expressed in Lahar (1996):

> Women and other oppressed categories of people should be especially wary of paradigms that could be constructed as advocating the sacrifice of individual needs to a "greater whole" – whether that be the family, society, or "Gaia," a planetary entity...[E]cofeminist theory must be especially careful in outlining its guiding principles to address the importance of individual beings *as well as* the larger whole.

The field of 'eco-feminism', and inevitably that of feminism generally, have consequently been transformed in a typical postmodern fashion within a much broader discourse. This broad discourse, as in many other fields, is however always characterised by a continuous caution about introducing any formal framework that could potentially limit and oppress individual expression.

This brings us to the last separation established during the modern era - the separation of people. It will be illusionary to believe that any new paradigm that will mobilise scientific knowledge, narrative knowledge and a new sense of spirituality can be achieved without serious efforts in overcoming the separation of people. In fact, this study is firmly based on the belief that this kind of mobilisation is only possible as an *emergent* process resulting from cooperation and collaboration of people. It is also clear that the diversity of human cultures, religions, and the specifics of their natural environment will inevitably lead to the emergence of many different paradigms, some very localised, and some possibly more widespread. In any case this process will only be realisable together with radical transformations of the present practices of decision making processes in our societies. There is already a significant awareness about the kinds of transformations that are needed, as exemplified in the continuously repeated requirements for 'public participation' and 'involvement of all stakeholders' in the decision making processes. However, regardless of the official recognition of the need for these transformations, it has remained extremely difficult to realise them in practice.

This study proposes that the processes of decision making can be transformed through the introduction of new kinds of technical means and institutional arrangements. All the subsequent chapters of this thesis will be devoted to some possible ways in which these new kinds of technical means and their institutions may be researched, analysed, designed and implemented in order to achieve this goal, albeit restricted to the fields of water and the environment. However, it needs to be stressed already that the kinds of technical tools and systems that will be proposed must always be regarded as means aimed at *transforming* the present practices of decision making, rather than supporting them in their existing form and practice.

This approach will then certainly clash with the present practices of decision making, which in most cases are carried out by so-called 'decision-makers' representing the interests of various corporate organisations. These centres of decision making are always either market or state institutions. Both the state and the market are of course legitimised in the name of individualism, but in reality they both in some sense limit and reduce our individual choices, taking away our own human responsibilities and in so doing leading us to neglect our individual interests as subjects. The market shapes us as free human beings, but this freedom is reduced to our freedom to consume, and the market usually proclaims this freedom as the ultimate one. At the same time, the state has more to do with taking the responsibilities for our actions, but in so doing it creates its own 'state ethics' which is very often far from satisfying our personal moral standards. Both these instruments "dwarf the most modern of the modern person's qualities: the ability to choose autonomously, and choose where it really counts" (Bauman, 1996, p.183).

In fact, the approach required to overcome the separation of people through the introduction of these new kinds of technical tools and social arrangements will inevitably be framed in our postmodern condition, frequently described as 'modernity without its illusions' and the already mentioned 'incredulity towards metanarratives'. The design and development of the proposed tools will therefore have to be oriented towards discouraging the introduction of any framework that inherently disregards the interests of individuals and groups of individuals gathered around a common concern: such a framework inevitably demands a 'sacrifice' of these interests in the name of some metanarrative about a "greater whole".

While we are witnessing the collapse of the separations introduced by the modern constitution, our yearning for an understanding of all these processes must cause us to turn to the extensive critique of the modern project as this has been so deeply expressed in phenomenologist and existentialist philosophies. Already from the middle of the 19th century, the dangers hidden in the paradoxical modern constitution were recognised and brilliantly expressed by thinkers such as Kierkegaard, Dostoevsky and Nietzsche. Their line of thought was later thoroughly expounded in different ways in the works of Husserl, Heidegger, Marcel, Sartre and others. All these thinkers, in one way or the other, have been struggling with the crucial, destructive 'force' that ultimately drives the modern project, that which for Heidegger "does not even know its own name". For Kierkegaard this was *Intet* (literally 'the nothing') as a sickness of the spirit but a sickness that was still essential to us. For theologians such as Barth, following a tradition that traces its lineage back to Augustine, this became *das Nichtige*. For existentialist philosophers such as Sartre, it was, similarly, *le néant*. English translators appologise for their inability to express this in any other way than by 'Nothingness'. Its individual and social manifestation is better known as *nihilism.* Although it presents itself in so many different forms, its origins are in one separation, so far unmentioned, but in a way underlying all the other modern separations. It is the *separation of knowledge from true wisdom and therefore faith*, and thus the separation through the 'downside' of the 'postmodern' condition or experience, from all our repositories of wisdom. The characterisation of modernity that directly reflects this condition is the famous pronouncement of Nietzsche that 'God is dead and we have killed him', even more exacerbated in Dostoevsky's words that 'if God is dead, everything is permitted'. Following Heidegger, the nominally 'non-religious' may prefer to understand the reference to God in these pronouncements as a reference to any suprasensory framework, which by virtue of providing grounds for human value systems becomes also an *obligatory* framework. The paradoxical modern constitution removed the possibility within Western culture of relying on a framework founded in Christian ethics and left an empty space, requiring this to be filled with a reason-justified alternative. This requirement turned out to be extremely difficult, and ultimately impossible, to fulfil, and indeed it is just as we pass from the modern into the postmodern condition that we gradually discover that it might indeed be unfulfilable, and that reason is not that 'almighty' after all. In the meantime, so many different candidates continuously occupy this empty space which was supposed to be filled with Reason: different ideologies, political or spiritual 'leaders', gurus of every kind and conviction, new nations, new classes, new political parties, 'new religions'. Some are indeed reason-justified, but most are not. But exactly this process, in turn, has became the key characteristic of modernity. It has, in effect, transformed the world of obligations into a world of options. Not *some* options, but an infinite number of options, *limitless* options.

But, if 'everything is permitted' than the 'kingdom of nothingness' appears to rule. As we will see in the following section, from one perspective our condition may also be described as *hypermodern*, as the powerful forces of modernity still continue to shape the key aspects of our lives. In this hypermodern era nihilism continues its rule in the form of a total reduction of all human values into one exchange-value. And this should not be understood only in the mercantile sense of buying and selling the self. It refers to the endless possibilities for conversions of the self, without any ground against which these conversions can be defined (Margolin, 1995). The modern man appears both to himself and to all the others as a mere assemblage of 'functions', regardless of whether these are vital or social functions (Marcel, 1956//1984//1995). Everything in the modern project has been done in the name of the individual (*in-dividual!*), yet the individual has been divided into an endless number of functions. It is noticeable how this same method that has allowed the reduction of the natural world into a pool of resources has been applied to reduce the whole of humanity into a collection of 'human resources'.

This situation is frequently described as the so-called 'death of the subject', a result of the modern objectification of everything, from nature to humanity. Both the philosophy of phenomenology and that of existentialism have contested this notion on the deepest experimental and theoretical grounds. Phenomenology insists that apart from the outer, 'objective' world, each person as a subject has his or her own 'inner world', and that 'reality' is no more than the name that we give to the interface between these two worlds (Abbott, 1994). Moreover, the first and supreme cognition of the outer world is, as mentioned earlier, always pre-predicative and pre-linguistic and occurs through *experiences* of *phenomena*. Husserl (1948//1973), the most prominent of all phenomenologist philosophers expressed this clearly:

> Objectification is always an active achievement of the ego...

This sentence strikes at the core of modern epistemology. It challenges the way in which the modern era postulated the possibility of acquiring knowledge. Even while the outer world truly and objectively exists, our cognition of it, the very process of *abstraction* that reveals to us the objects within this world with their properties and relations, is always a product of subjective *experience*.

Existentialist philosophy goes even further in its efforts to reclaim the depreciated position of the subject. It breaks completely with the ontology of classical philosophy of 'essence before existence', and claims that the human mind itself works the other way around. There are no *essences* in the world 'waiting' to be discovered by Reason. Instead, these are just concepts that are abstracted from the flow of our life experiences. Human existences come first, and essences are continuously being created. But this position, in turn, brings one crucial consequence. If we indeed *create* essences, ontological categories, and ethical values, this puts an incredible burden of responsibility upon us. Moreover, human *creativity* acquires a value of crucial importance, as it remains our only capacity to contest the forces of nihilism. Sartre nicely expressed this in his pronouncement that 'if God is dead, everything is indeed permitted, but nothing can be forgiven either'. More precisely, it is through creativity that the person transcends his or her self, coming into contact and drawing upon other forces that are not personal, even as they are not at all impersonal. Some would speak of 'inspiration', while the more religiously aware would speak of 'grace', but to refuse either when it is offered is unforgivable.

These philosophies offer brilliant insights into the deep problems that humanity faces in the modern era. It is not difficult to trace the many important influences that they had on a body of postmodernist thought that further challenges many crucial modern concepts. On the other hand, regardless of the incredible power that they have in this respect, they have had only a limited effect upon the 'modern' project itself. Thus it has become common to identify the notion of a 'European way' towards building social relations, as embodied in the *Social Charter* of the European Union, as opposed to an 'Anglo-Saxon way', as a natural consequence of these influences. There is almost certainly some residue of truth in this, no matter how opportunistically this is misused within the European political arena. Otherwise, however, these understandings have been largely impotent in actually changing the direction or the pace of modern development. Such was the power of the modern constitution that virtually all challenges on theoretical grounds could be simply ignored. The fact that we are able to see so much more of the illusive and paradoxical nature of the modern separations is only because we are ourselves confronted on a day-to-day basis with the *real-world* events themselves, as they proceed as parts of the practice of modern development. It is only then that we turn to these deep thinkers in order to arrive at some understanding of the often terrible events proceeding outside of us, and deep within our own selves as well.

The real awareness of the necessity to cancel out the separations came only after the growth of the 'hybrids' had become so significant, involving stakes related to the very survival of our own selves and our entire planet. It has taken environmental disasters proceeding at enormous scales to make us realise that some kind of 'environmental ethics' is needed. At the same time it has been necessary to reach the stage where it becomes clear that the sustainability of human society is practically impossible without employing our artefacts. It only thereby became possible to start thinking about a 'sociology of things' and a 'social psychology of objects'. We suddenly realise that the way in which we are approaching science, technology and knowledge in general is still lagging behind the reality of our present time. The 'hybrids' are therefore coming to be recognised more and more clearly, and the separations between nature and society, as well as between 'scientific' knowledge and 'narrative' knowledge, is being cancelled out. With respect to these separations, the realisation is already emerging that "boundaries [created by separation] only stabilise when they cease to be stakes in the game" (Lyotard, 1979//1984, p.17). Even when looking from that side alone, the concept of sociotechnology already imposes itself as the only approach appropriate for dealing with our current problems.

This process on the other hand leads us to re-examine the two other key parts of the modern constitution, the separations between people, and the separation of 'man from God'. These separations remain more concealed, and it needs some effort to discover that they have to be challenged as well. The way in which these challenges can be put appears even less clear. On one hand, the pre-modern, pre-scientific world is gone forever and has passed beyond recall. Very few people are prepared to sacrifice any of the liberating concepts that emerged with modernity, and certainly not for any kind of 'greater whole'. Nobody will give up human rights, scientific control over the environment, wellbeing in this world as a desirable goal, or confidence in one's own ability to do something entirely different and significant in this world (Gallagher, 1998). Yet, at the same time, the awareness is spreading that many attitudes, both towards our fellow human beings and towards our natural environment must be radically changed. In that sense we will have to understand that sociotechnology as a concept means very little without the *active* participation of people. Moreover, the benefits of this concept can only be harvested if in these processes people are approached as *subjects*. This key realisation should discourage their recasting simply into *stakeholders* and a *general public* with *interests* that must be taken into account. Instead, they need to be recognised and genuinely accepted both as individuals capable of their own moral judgement *and* subjects capable of creative thought and action. Marcel (1956//1984//1995) described these kinds of relations among people by introducing the term *intersubjectivity* into this debate (see also Davignon, 1985 and Klinkenberg, 1996, p.125). We would like to propose here that the concept of sociotechnology is inseparable from - and indeed meaningless without - intersubjectivity. In fact, the kinds of systems that will be the main subject of this study must always be considered as the '*principal sociotechnical means for promoting intersubjectivity*' (Abbott, 1998). Thus:

Thesis 4: *Our tools and institutions will have to work together to raise the state of aporia that has been induced by the modern constitution. These sociotechnical constructs have to interact with other awareness raising processes in helping individuals to change their attitudes towards their societies and towards nature. This means, to adopt another expression of Gabriel Marcel, that they have to encourage individuals to promote their powers of* second-order reflection, *even while necessarily interacting with other individuals. These constructs must then function as devices for promoting intersubjectivity.*

2.5 The postmodern and the hypermodern

In order to understand how sociotechnology critically depends on the active involvement and participation of *people*, it has to be further identified as a *postmodern* concept, regardless of the notorious difficulties associated with the meaning of the term – 'postmodern'. Most of these difficulties arise from the fact that 'the postmodern' has become a central theme in so many different domains of human thought and activity that it is now seemingly impossible to trace a common thread which connects all of them, except for the word 'postmodern' itself. The term was originally introduced in relation to architecture, pictorial art, and literature, but soon it became central for the whole field of aesthetics. Politics, psychology, sociology, economics, and even law, science and philosophy itself, have been transformed in the last few decades of the twentieth century under the powerful influence of postmodern thought. This penetration has been so strong that our contemporary culture as a whole is most often nowadays described as a postmodern one[4].

[4] It is significant to note that in fact the technological fields, especially engineering, have been the least influenced by postmodern ideas. While for contemporary art, and architecture in particular, the postmodern 'wave' has already been largely assimilated, technological fields are only beginning to realise their new tasks and responsibilities in our postmodern condition. As at many times in human history, this will be the key condition for new kinds of technological developments, inspired by art and conditioned by society, that will surely bring new breakthroughs and will ultimately radically change a way of life.

In the attempt to extract some meaning from the innumerable, often confusing and contradictory descriptions of our postmodern condition it is perhaps helpful to follow the distinction made by many authors (see for example Botella, 1995), between three different uses of the term postmodern. *Postmodernity* is the term that describes the social and historical condition of our present age. *Postmodernism* is the term that refers to the cultural and especially artistic expressions of this condition. *Postmodern thought* refers to the underlying philosophical and theoretical concepts that have emerged primarily as a challenge to the basic concepts of the 'modern project', rather than as foundation for some new 'project', which might be called a 'postmodern project'. Very often the terms have been used interchangeably, and in many cases this is justifiable, since the three terms refer to interwoven aspects of our present condition. We propose this distinction here mainly for two reasons. The first one is clarity of presentation, and the second, as will be shown later, is the frequent misuse and intentional inversion of the meanings of the terms postmodernity and postmodernism within the still powerful and even more omnipresent (hyper)modern discourse. Therefore we shall first try to concentrate on the third term – postmodern thought- and try to present its key concepts.

Postmodern thought, through its theory, makes an at first sight impossible attempt to undermine the ingenious, because paradoxical, modern constitution. In order to do this, it challenges the central position of modernity, that reason, and in particular individual rationality, is the one and only firm ground for all modern theories, and, for that matter, the whole notion of 'progress'. Postmodern thought, however, does not attempt to offer another alternative to individual rationality, but questions the very possibility of having a ground that will offer a proper foundation for any such 'totalitarian' theory. In that sense, even the expression, postmodern *theory*, is arguably meaningless in many fields of application when seen from the postmodern point of view. There are correspondingly several concepts that may appear as central in postmodern thought. Each of them taken individually has perhaps insufficient power to change the established 'mindset', but their union has offered enough intellectual momentum to form an alternative worldview to that of modernity. Let us try to explain the key postmodern concepts by adapting again some notions of Botella (1995). As we shall see, they are all in one way or the other reactions to the recognition of the limits of modern *epistemology*:

Groundlessness is, as we have already stated, the central concept in postmodern thought. Rational inquiry has no more any specially privileged access to reality, regardless of how reality is defined. In fact, there is no epistemic foundation that will by itself give a privileged position to this or that claim of knowledge.

Situatedness (or *context dependence*). This concept is rightfully linked to the position that *universal* knowledge is an impossibility. All knowledge is *situated* in a given context, and any claims about something being valid or even 'true' should be restricted to local and specific circumstances. Claims for generalisations with a tendency to move towards universality should be avoided and extensions beyond the given context should be made with extreme caution.

Constructivism is a key concept, although it is related to groundlessness. Instead of rationally justified knowledge as that which leads to the discovery of truths about an objective reality, knowledge now becomes little more than a structure of human constructions. At the same time, however, the key adjective that is added to this concept is *social* constructivism. It indicates that the process of constructing knowledge and its articulation in *language* is a result of social acts, and it does not happen in the isolated minds of individual rational agents (as predominantly maintained by 'modern' epistemology).

It is obvious that the central preoccupation of postmodern theory is the question of knowledge: the possibility of knowledge, its production and validation, and above all, as we shall see, the question of *legitimation* of knowledge. The *method* for approaching the questions of knowledge, however, is here entirely different as compared to the modern 'rational' or 'scientific' method, and then exactly because of the groundlessness and the rejection of the supreme position of rational inquiry. Instead of dealing directly with ontological or epistemological categories, asking questions about what objectively exists and is to be discovered, and what is subjectively constructed, postmodern thought takes the so called 'language turn', often described as the 'descent into discourse'. (The most influential of all varieties of this method is certainly Derrida's method of *deconstruction*, which forms the cornerstone of what is known as 'postmodern philosophy'. See, for an introduction to this, Powell, 1997; Norris and Benjamin, 1988; or Neal, 1995).

This method studies the pragmatics of language and the corresponding 'language games' that arise in any given discourse or social context. However, language is not studied as a 'mirror of the world'. It is not just an efficient means of *representations* which can be potentially shared and therefore lead to a common agreement about any knowledge or even about meanings of such firmly established notions as 'reality', 'truth' or 'objective view'. Postmodern theory's 'language turn' is in some senses exactly the opposite: language gains its meaning through various kinds of interactions among people, and from the ways in which language is used. 'Truth' and 'reality' are conditioned on the participation of people in a set of socially constructed conventions: by a shared 'inner world' as well as a shared 'outer world'. An 'objective view' collapses into 'playing by the rules' that are emerging from the human interactions. Universality disappears, and all these notions become inevitably 'localised' and restricted within the boundaries of a given discourse. So the claim that we can rely on any universal concept, and with that have access to a universal knowledge, vanishes, strictly speaking because there is no 'universal language'.

It has to be remembered, however, that language has been the primary focus of various schools of epistemological thought, for different projects and with different aims, throughout the late 19th and the 20th centuries. A focus on language is not at all an innovation of postmodern theory. The beginning of the twentieth century was marked by the attempt of the school of logical positivism to show how all knowledge could be *represented* in language by using the formal notions of logic and mathematics. This was a *truly modern* attempt that took language as *representation*. The works of the logical positivists (Russell, Whitehead, the early Wittgenstein and many members and followers of the so-called Vienna Circle) were all aimed at universalising the modern epistemology of rational-scientific knowledge through the use of formal language. The final goal was not achieved, but were made immense contributions to the developments of modern logics, mathematics and philosophy.

Another significant school of thought focused on language was the *structuralist* movement (initiated by the teachings of de Saussure). It studied language as a *system* with its own rules of signification that provide meanings. It was an attempt to approach language by 'excluding' the 'language user', and by revealing the rules of signification, that is, to somehow *objectify* the linguistic system. From our present perspective, it can be considered as another unsuccessful (perhaps even 'hyper-rational') attempt directed towards universalisation, disregarding the influences of contexts, individual and collective language users, and their *motivations* in the formations of a given discourse. Nevertheless it provided a fertile ground for the development of *semiotics*, the study of signs and of signification in general, which nowadays takes as its study virtually all aspects of human culture.

It is probably the philosophy of the later Wittgenstein (e.g. 1953//1992) which is closest to the way in which postmodern theory studies language, as has been recognised by many authors. Language has ceased to be seen as a primarily representational device and a sort of descriptor of human action, but has itself become a *form of action*. Meaning is not derived from a correspondence of language terms to real or imaginary objects, but from the *use* of these terms in interactions among people. At the same time, the social perspective becomes crucial. Individuals are never 'free' in this process of 'assigning meanings', but limited, and exactly by the 'publicity' of the rules resulting from their own interactions. Philosophy in general, epistemology and, in particular, the philosophy of mind, have been immensely influenced by Wittgenstein's work. Much of the present day discussions (and research in various fields) about 'socially situated knowledge' or 'socially conditioned intelligence' is profoundly influenced by his philosophy, which dismantled the priority of the 'first-person case'[5].

The 'language turn', that is commonly attributed to the school of postmodern thought is thus not something that came without a strong background of 'language studies', itself evolved through a long intellectual struggle. Careful studies will also find the origins of the postmodern concepts in the wider modern critique expressed in phenomenological and existential philosophies, even though these were not so much focused on studies of language. Postmodern thought has to be regarded as a development that was *conditioned* by modernity and its critique.

What does, however, the postmodern method achieve? The most important of its results is its capacity to *reveal* what is hidden in any discourse (and therefore in any social context), 'behind' language. The primary target of the postmodern method is of course the discourse of modernity, and particularly the ways in which within this discourse relations of dominance are constantly being established and maintained. Especially Derrida showed that by a careful dismantling of the 'text' (discourse in the broadest sense) it is possible to discover the underlying, very often deeply hidden, concepts, such as false dichotomies for example, that are a result of ideological, social, cultural or professional (mis)conceptions. What appears as a result of rational inquiry, and what is presented as an unbreakable structure by virtue of the

[5] Many other paradigms that now appear in various fields, ranging from computer science to sociology (agent orientation, multi-agent systems, emergent systems, complex adaptive systems, self-organising and autopoietic systems, etc) some of which will be the subject of Chapter 5, owe very much to this shift in view, even though it is rare to find references to Wittgenstein's work or 'postmodern thought' in the corresponding literature.

involvement of this transcendent 'free' rationality, is suddenly shown to be a cultural product, and because of its socially constituted character its applicability is shown to be really quite severely restricted. In fact the analysis goes much further, in showing how very often the discourse itself is used for silencing various voices in society, frequently those of the marginalised and the underprivileged. Moreover, the adoption of the discourse itself creates and sustains this marginalisation of others. It is therefore no surprise that postmodern thought had such a powerful impact in feminist, environmentalist, and multicultural studies. The whole postmodern approach comes to be seen by many nowadays as truly *emancipating*.

As a result, postmodern thought presents a world with alternative realities, representations, worldviews, and even 'rationalities'. This plurality rejects the 'monologue' of scientific-rational knowledge, and directly attacks its refusal to engage in any kind of dialogue with imprecise, inconsistent, sometimes even contradictory and absurd 'forms' of language. Indeed, these 'forms' of language are now raised up to a more equal level with scientific knowledge and are accepted as 'knowledge' as well. (It is characteristic of many postmodern authors to use the word knowledge*s* deliberately – in plural[6] - exactly to make this point even more clear (as, for example, in Gergen and Thatchenkery, 1997). As we have already mentioned, Lyotard used the term 'narrative' knowledge in order to denote this same distinction.

At the same time, in parallel with this development, a very strong opposition to and critique of postmodern thought has developed which is also significant for the understanding of our postmodern condition. In one way or the other, all critique of postmodern thought is about one central question: the problem of the *legitimation of knowledge* (for an excellent presentation of this problem, see again Lyotard, 1979//1984). The key objection to postmodern thought and its method is that it appears to be applicable to the subversion of all kinds of established structures of knowledge, and indeed to relativising almost any statement or knowledge claim. This results in a strange paradox. The 'universal statement that no universal statements can be made' then appears to leave no room for any meaningful human action. If, after all, there is no ground on which this or that epistemic alternative can be justified, where do we find the motivation for action? Critics of postmodern thought (e.g. Norris, 1993) argue that this kind of epistemological relativism is not emancipating at all and that it actually leads into a 'blind alley' where 'anything can be said about everything'. Moreover if there is any attempt to provide such a justification for any alternative, what will prevent us from treating it as yet another conveniently 'constructed' rhetoric, as the method apparently allows us to do? Because of this impossibility to *choose* between this or that epistemic alternative, postmodern thought is accused of becoming the opposite of what it promises. It is.accordingly accused of spreading defeatism and ambivalence, leading to a total loss of faith and a universal hopelessness that overwhelm the individual to such an extent that no meaningful action by that individual is possible.

---

[6] The reader should be warned however, that the same 'trick', the use of the word knowledge*s* – in plural - is also used by authors to refer to the innumerable narrow specialisations that are the result of the modern compartmentalisation of rational-scientific knowledge (e.g. Drucker, 1994). It should be obvious that this is not the intended use of the word here. The same word is actually used by proponents of radically opposing views on the new nature of knowledge.

When postmodern thought is faced with this criticism, it encounters an extreme difficulty in defending its position. Probably the worst possible reaction by many proponents of postmodern thought has been the complete 'escape' form this criticism by enclosing their discussion within the elite 'safe houses' of academia, with a rhetoric that was frequently removed beyond any comprehension. The whole postmodern programme became of interest only to a small coterie of a couple of thousands scholars, without any relation to persons actually living in 'real-world' society. The resulting discrediting of postmodern thought (and even sometimes the pejorative use of the term) is certainly more due to the complete abandonment of any attempts to have a significant impact on events in the real world, then to the incomprehensibility of the rhetoric. Of course, this situation has provided even stronger arguments to the critics of postmodern thought.

On the other hand, many thinkers who felt that postmodern thought had to face its critics openly, found that this could only be achieved by introducing additional concepts over and above the three that we have already introduced. Many have agreed that the one additional concept that might provide an exit from the epistemological 'blind alley' is that of a so-called *neo-pragmatism* (Polkinghorne, in Botella, 1995). This concept relates any body of knowledge only to certain *tasks* arising immediately in a given context. These tasks determine all the tests for the viability, coherence and the predictive power of the knowledge claims. The focus here shifts completely from the truthfulness of knowledge to its *usefulness*. Although this appears as a solution to the problem of relativism, it still leaves open the key questions: Who decides what the tasks are, and who decides how usefulness is to be defined?

It seems that any attempt at answering these questions necessitates a stepping out from the closed shell of theoretical discourse and a looking at what actually happens in 'real life'. In order to understand this, it is necessary to come back to the distinction made between the various uses of the term 'postmodern' at the beginning of this section. Postmodern thought is not the same as postmodernity and postmodernism. Theoretical discourse always remains isolated and powerless unless its interaction with real-world practice is brought into question. Postmodernity and postmodernism are related to the successful development of certain theories, but only to a limited degree. A complex mixture of social, technical and historical developments during the 20th century are arguably more responsible for the raising of serious doubts about the promises of the modern project. The catastrophes of the two world wars, the subsequent cold war, and the growing awareness of the scales of environmental destruction due to uncontrolled agriculture, logging and heavy industrialisation, have had a direct and powerful influence on the creation of postmodernity. At the same time, the explosive development of the mass media and the information and communication technologies have brought about a widespread awareness that there are indeed many different worldviews, and that it becomes increasingly difficult to justify the supreme position of any one of these over all the others. *Diversity* has entered the scene in virtually all possible domains of human thought and action. Cultural diversity has became not only respected, but also recognised and cherished for its sustaining role in the new 'global village', in a manner similar to the recognition that biodiversity is the key factor for preserving our precious ecosystems. Epistemological diversity appears then as a consequence of this whole process, and not as a cause.

In the course of these developments, the two principal instruments of modernity, the market and the democratic nation-state are also being transformed, and especially in their mutual relation. The postmodern condition destabilises the 'traditional' position of the state as the ultimate social unit where both the individual and the collective needs and rights of its citizens can be satisfied and guaranteed through state institutions. A new kind of consciousness has been created, that we are all 'citizens of one planet' and therefore inevitably subject to a common destiny, and, at the same time, that we are a conglomerate of diverse cultures to an extent that different kinds of social arrangements are unavoidable, and perhaps even recommendable. This opens the possibilities for the evolution of different kinds of social entities that in some respects can be far superior to those of the nation state. The innumerable non-governmental organisations, regional and supranational bodies in various domains (economic, military, cultural, etc.), 'networks' of people and organisations (both real and virtual), are all in one way or another taking up domains which were traditionally reserved for the state's own power and responsibility.

At the same time, this 'crisis' of the state is being extremely successfully used by the market. In fact the whole postmodern condition, characterised as it is by plurality and diversity, has been embraced as well as generated by the market. The crucial point is that knowledge does not escape from this process. Quite to the contrary: knowledge itself is turned into a *commodity*, and becomes increasingly produced to be bought and sold. In fact, the question of the criteria for legitimation of knowledge, left open by the postmodern postulate of epistemological diversity is now answered using the criteria of the market. In a typically postmodern fashion, we no longer ask the question: "What is true?". The neo-pragmatic condition advises instead to pose the question of "What is useful?" When faced with the doubts left by the neo-pragmatic condition about the definitions of tasks and usefulness, the market provides the ultimate, and seemingly the only relevant question: "Can it be sold?" The criteria of *performativity* and *efficiency* are re-introduced as the key principles for judging what knowledge is, and how it is to be valued.

The proponents of the present day *laisser-faire* economy, of course, present this as the only 'natural', and thereby 'inevitable', development[7]. In fact, the postmodern condition has already shown that there are many different ways of 'knowing the world'. Consequently, different, and possibly even new kinds of social organisations are then possible, at least in principle. This is however perceived as a serious threat of destabilisation by many established 'orders' in society. A counterattack by the corporate elites in defence of 'modernity' and 'progress' has consequently been triggered. In the new, postmodern reality, however, this elite has had to reintroduce a new, or at least different kind of modernity. This aims to re-elevate the position of reason, but now within in a new context, and one that accepts epistemological diversity. As expressed by Dickson and Noble, (in Said, 1985) this elite has managed to "proclaim a new age of reason while *re-mystifying* reality". This has been done

[7] It is also however the (mis)use of the *rhetoric* of 'laisser-faire' economy, and not its reality. New kinds of interventionist and protectionist measures are being introduced continuously in today's world economy, with the difference that the framework of institutions within which this is being carried out is changing. For example, nation states loose certain of their trading power to institutions organised around 'multinational trade agreements', such as World Trade Organisation (WTO), for example.

by introducing new kinds of reason-justified epistemological and ideological 'imperatives' (ibid):

> The rediscovery of the self-regulating market, the wonders of the enterprise, and the classical liberal attack on government regulation of the economy, all in the name of 'liberty'.
> The reinvention of the idea of progress, now cast in terms of "innovation" and "reindustrialisation", and the limitation of expectations and social welfare in a quest for productivity.
> The attack on democracy, in the name of "efficiency", "manageability", "governability", "rationality", and "competence".
> The re-mystification of science through the promotion of formalised decision methodologies, the restoration of the authority of expertise, and the renewed use of science as a legitimation of social policy through deepening industry ties to universities and other "free" institutions, [including those] of policy analysis and recommendation.

These 'new' imperatives set the stage for the alternative 'true' nature of our present condition: *hypermodernity*. Why make this distinction between postmodernity and hypermodernity? The crux of the matter of nomenclature is that the reshaped instruments of modernity have absorbed many ideas that emerged within the postmodern condition which were originally, and are even today, strongly attacking the core of the modern world-view. Through this process, the term 'postmodern' is itself subverted, being imbued with a new meaning. It is suddenly taken to refer to a cacophony of dissonant voices, with the appearance of celebrating differences among people, ideas and cultures, where in fact it celebrates the logic, power and inevitability of a late hypermodern capitalism (Jamison, 1985, 1991). Hence arises the association of postmodern ideas with such negative cultural realities as the eclectic postmodernism characterised by the slogan that 'anything goes' or the various anti-modernist strands which grow together to form a new neo-conservatism. These 'false' brands of postmodernism, as Lyotard (1979//1984) names them, tend to present the postmodernist project as something which is not critical, oppositional or contestary to the logic of modern consumerist society, but only contributes further to the reinforcement and replication of this form of society.

The way in which modernity re-asserts its power is not new of course. Several rebellions against the 'curse of modernity' have occurred since the end of the 19th and throughout the 20th century, but most of the attempts to bring these reforming ideas into practice were ultimately 'cursed' with the taint of modernity. This process is probably easiest to perceive in the developments in the arts throughout the last 100 years, where each rebellion against the previously perceived 'modern' art has been immediately 'modernised' (or we could say 'tamed' or 'absorbed' or 'assimilated') within the social structure governed by the rules of the critics and the art market. There is scarcely any single artistic attempt to contest modernity, no matter how revolutionary and extreme in its nature, that has not been taken over by the establishment of the art critics, art dealers, and, generally, by the market in order to convert its achievements into commodities for sale. The way in which 'the market' does this, by subverting the entire discourse by degrading this into mere *Gerede* (gossip, chatter and idle talk) was masterfully described by Heidegger (1927//1962 pp.210-224)

Although many thinkers even in the last but especially in the present century, have foreseen the spread of this condition, there does not seem to be any straightforward remedy. (The word 'remedy' fits in well with the description of our present condition by many postmodern thinkers as it suggests some sort of illness - 'historical amnesia' (Jameson, 1985) or 'a new form of schizophrenia' (Baudrillard, 1985). The 'radical humanisation' of society is definitely erased as an alternative. Introducing legal measures to control or reduce the so-perceived crucial "growth" is impossible because it would clearly bring disastrous consequences. At the same time, none of the postmodern ideas, whether about humanisation, plurality, nostalgic reminders of a lost touch with the 'real' outer world, or even the contempt for consumerism, can escape the apparently omnipotent market forces. Through a correspondingly sophisticated advertising enterprise, all these ideas are *themselves* again turned into commodities for sale. Nowadays one can see and hear advertisements with slogans like "connecting people" (global telephone network company), "image is nothing - thirst is everything" (soft-drinks-producing company), or advertisements using songs like: "beautiful people... you live in the same world as I do...I did not notice you... I am ashamed to say..." (Insurance company!). A weekly magazine advertises itself on television by 'taking' celebrities into the open spaces of Antarctica, using background projection of course, from where they send messages about how these open spaces make them humble and from where they can call upon their politicians to share this same experience. Oil producing companies advertise themselves as advocates of environmental protection and care, leaders in the quest for renewable and clean energy sources, and promoters of partnership with third world societies. Perfume producing company advertises its latest product with the slogan: "against the bad odours of capitalism…"!!

Because Internet, and electronic networks in general are so determinant for this study, it is important to stress the special role of the electronic media in this hypermodern picture. When taken over by the market, Internet along with all the other technical developments which are building our Cyberspace (such as a 'classical' television now made even more powerful by cable network connections and satellite coverages) are contributing strongly to establishing the profile of 'the human' as the one who is *free to consume*. The illusion that man is free in this hypermodern society is nowadays increasingly maintained by the existence of two decreasingly intersecting outer worlds. The one of these is the 'real' one of direct sensual experience, which becomes more and more distant, while the other is created by Virtual Reality techno-media simulacra. In this second outer world, in contrast to the first, men and women feel themselves secure, and there are no real uncertainties. They can choose their own experiences, which thereby become life experiences, so as to act out any fantasy, and their only limitations are the amount of money that they can afford to spend in order to realise this state. One manifestation of this movement towards virtual realities is the immense growth in the computer game market, now itself increasingly connected to Internet as a means of constantly 'refreshing' the game content.

As the existentialists in particular so often emphasise, one of the key differences between we humans and our fellow creatures is that we lack much of the perfection of the animal instincts and that we use our consciousness instead to survive in our outer world. However the price for possessing such a developed consciousness is that we have to live with uncertainty and doubt within our inner worlds. This is probably one of the heaviest burdens that humans carry throughout their lives and they very often look for means to relax and avoid it, even though many would accept that it is only by coping with this burden that we become truly human. It is this uncertainty, doubt and ultimately tormenting ambiguity which makes us human, with all our human responsibilities, and it is at the origin of our moral selves (see again Bauman, 1996). We are all happier if we have some guideline and justification for our conduct. It is even easier for us to make wrong decision while still being able to point somewhere else for our excuses than to make a right decision without having anything with which to justify ourselves. For a long time this security has been provided to man exclusively by God, and it still is for many people in the world. Modernity has however apparently opened the way for many possible alternatives. The hypermodern era thus appears to provide something entirely different. This is the possibility to live an isolated life in an own private Cyberworld where the discomfort of uncertainty is present, but where all we now have to do is to 'press a button' for this to be dispelled. The 'real' world at the same time becomes dematerialised, distant and less important. Reality, as the interface between our inner and our outer worlds, itself becomes 'destabilised'. The market has taken over the most important condition for our peaceful life, in that it is this which creates the illusion that we are in some sense 'safe'. It is now more and more the market that determines what is 'true' and what is 'false' within its virtual worlds.

The transition from a modern into a postmodern condition has resulted in a rejection of the supreme position of reason over desires and instincts, and in some sense it has provided a liberation from the feeling of resentment towards civilisation and towards the self. Cyberworld offers a new realm where the suppressed desires can be now unleashed unbounded, without taking true reality into account at all. The attractiveness of this realm is even more increased since the entrance into, or the exit from, this virtual world is only one click away. Once the computer, the TV set, the play station or the video recorder is switched off, true reality reasserts itself, and with it the despair of not being able to "choose where it really counts" is re-awakened. As in a feedback loop, the hypermodern individual is drawn back to the simulacrum, since this is the only place where he or she feels safe and satisfied. Through this process, reality indeed becomes more and more displaced and the very sense of reality of the individual is gradually eroded, and then in the sense that it does not 'count' even if it is there. The boundary between a simulated, virtual reality and a true reality becomes ever more often *masked* rather then *marked* (Margolin, 1995). This is how we actually enter Baudrillard's 'hyperreality' which has no reference to any experience outside of its own constructions, where the simulacrum replaces the real completely.

This form of hypnotic state is an entirely new condition for humanity. In many ways it is impossible to foresee its consequences, because there is no historical precedent at all which may serve as a source of experience. We can however already point towards some aspects that are becoming abundantly clear. Firstly, it becomes increasingly difficult to escape from this virtual world. The simulacrum of images seems to be everywhere. This age, which is often described as the one marked by the 'death of distance' (Cairncross, 1997), has not only managed somehow to 'kill' all distance, but leaves no possibility to take a *critical distance*, indispensable to all processes of understanding. Secondly (and even more disturbing), by staging all reality through the sophisticated media apparatus in a form of simulacra, our own personal morality is in a condition to which we are not adjusted. Individual morality, by 'nature' is essentially bound up with *proximity and closeness*. Its universe consists of family, friends, colleagues, neighbours, contemporaries, and places that we know personally, and for which we have a sense of belonging. Outside this realm, the modern 'state ethics' takes its own responsibilities, even if its standards are often so discomforting to our own individual morality. One would expect that through the 'death of distance' the universe of individual morality would itself be enlarged, and that places, people, and actions which are distant in space and time would become a closer part of it. We have at least come gradually to realise however the desperate *need for such morality at a distance* in a world where human actions have such far-reaching consequences. However, if through our electronic windows we see everything in our world as *a staged event,* happening *somewhere else*, our morality remains untouched and undisturbed. We have stopped seeing *real* people and places, and observe instead only 'streams of events' and 'unfolding stories as they develop'. Instead of developing *morality at a distance*, our moral responsibility becomes even more dissipated. This is probably the darkest side of hypermodernity, as it brings this kind of sophisticated adjustment and acceptance of injustice that destroys all motivation for action, and it is this that is probably a much more important reason for the feelings of hopelessness, loss of faith and ambivalence, which are wrongly attributed to postmodernity.

This brings us back to the promises of the postmodern thought and to its critics. The crucial mistake made by the critics of postmodern thought is to regard it as a development outside of its social, cultural and historic conditions, and thus to require of it that it provides certain *guidelines* for human thought and action. A body of thought which destabilises epistemological grounds and demands freedom from any metanarrative guidance cannot be expected to offer any such guidance. Invoking additional criteria in postmodern thought for defending it against its critics transforms it into yet another metanarrative, and it destroys the very postmodern spirit. The neo-pragmatic criteria offered for this purpose can in fact have far reaching consequences in that direction when opportunistically (mis)used through the market. The critics' attack on postmodern theory is in fact 'modern' in nature, and misses the crucial point, so often neglected when our postmodern condition is discussed. Postmodern 'texts' that recognise how plurality and epistemological diversity are subjected to market criteria are not celebrating this condition (Lyotard (1979//1984) himself clearly identified this condition as an actual 'delegitimation' of knowledge). They are only a recognition of the facts, and are not statements of *value.* As a reflection of actual social and cultural conditions they do not offer solutions to the maladies of our societies, but are more directed to identifying their symptoms. Postmodern theory does not ask people to *choose* between different epistemological alternatives, face the problem of justification of their choice, and be forced to accept the market criteria for knowledge legitimation. Instead, people are recognised as key subjects that can *create, develop, and change* these alternatives through their own, individual, active participation.

It is in this sense that sociotechnology is a truly postmodern concept that can open a new path for this kind of pro-activeness, and participation. It is a kind of postmodern technology, with new tasks and responsibilities that can be achieved only if it becomes a true sociotechnology. This is then a technology that sees itself as embedded in a given social context, recognises the crucial role of people in this context, and avoids presenting 'ultimate solutions'. Technologists, designers and engineers are also part of this context and their active participation actually transforms the whole attitude towards technological design. Their role becomes crucial in avoiding the closed loops of 'technological fixes' (and sometimes precisely by realising that a necessary preliminary to the activity of technological design is accepting that what needs to be done is to stop designing!). The power of modern technology makes the acceptance of the concept of sociotechnology among technologists themselves more urgent than among any other professional group. The escape from the hypermodern grip in which technology has been placed must be found through this new kind of practice (Margolin, 1995):

> ...the more a designer or an engineer can conceive of a user as a person of depth and worth, the more likely he or she is to design a valuable product...As designers and technologists develop a more *caring* feeling for how people live, they may also generate new products that respond to previously unimagined human activities. A discourse on human purpose generated independently of the market is not utopian.; it can have an effect on what the market produces.

The new nature of sociotechnical design will however face one problem of typically postmodern character: the absence of 'rules' and 'recipes' that will guide this activity, and upon which one can rely to end up with a 'good product'. "Postmodernism means working without rules in order to find out the rules of what you have done"(Appiganesi and Garratt, 1995). It means constant experimentation and openness to all influences, even without knowing where the experimentation will lead. Again, this must not be confused with the frequent misunderstanding that this presupposes a rejection of all (or even any) previous knowledge. The repeated calls stressing the need for 'continuous education' and 'lifelong learning' for example, are closely linked with a sociotechnical design, but are nowadays too often misunderstood at exactly this point. The success of this endeavour relies on exactly the opposite - mobilisation of *all* available knowledge – scientific and narrative, social and technical. The question of *how* to achieve this mobilisation requires understanding of the dual and apparently contradictory postmodern/hypermodern condition of knowledge (recognition of 'knowledge*s*' versus 'knowledge as a commodity'). This understanding now becomes critical for the sociotechnical approach.

Thesis 5: *The task of connecting people to a true or actual reality rather than to a simulated but false pseudo-reality is greatly helped by using much the same technical means as have been used to subvert true or genuine reality, namely the various simulacra. For example, the video game, as supported by an increased participation enabled by the Internet, and by Internet content 'refreshment', can be used to bring its players 'back down to Earth'. Moreover, the market itself provides some means to realise this project. In this sense the technology is 'neutral': it can just as well be used in the one genuine postmodern direction as it is now being used in the other, hypermodern, direction.*

2.6 The new nature of knowledge

The increased use of the word 'knowledge' for characterising all aspects of our present condition ('knowledge economy', 'knowledge society', 'knowledge age', etc.), may appear as yet another (and probably even more) short-term fashion, comparable to many other similar examples in the recent past (consumerist age, space age, information age, digital age, media age, etc.). Even if the fashion passes, however, the shift in the way knowledge is understood, produced, distributed and legitimised remains the genuine cornerstone of our present condition. There is indeed a new perception and use of knowledge, characterised both by the 'commodification' of knowledge and by the recognition of 'knowledge*s*'. The two aspects, as we shall see, are not necessarily contradictory, and there exists a certain compatibility between them that is particularly well suited to the sociotechnological approach that is pursued here.

As we have seen, the modern constitution initially gave knowledge the role of discovering *truths*, so that all *resources* could be better managed. Scientific knowledge, as the most successful of the enterprises engaged in this task, was either legitimised as necessary for the 'emancipation of humanity' or legitimised within itself as a 'speculative unity of all knowledge'. During the course of the modern development, characterised by the objectification of everything for the purpose of creating *resources*, a kind of 'meta-view' of knowledge has been created from which knowledge *itself* became understood as a *resource*. It became yet another resource, in addition to all the other resources (natural resources, later human resources) even if its application has been orientated towards managing all the others. With rational-scientific knowledge becoming the supreme knowledge, reigning over any other knowledge, the possession of it soon became the most important resource towards which any individual or organisation should strive. This was however a resource that was difficult to acquire and transmit. Education required long period of study and generally was considered as very expensive. Because of this, once individuals or organisations acquired knowledge they tended to keep it and were reluctant to share it. The *difference* in knowledge between the knowers, and those who lacked knowledge was the principle way to gain 'competitive advantage'. The social value of knowledge to those who could acquire, use and provide this knowledge became dependent on creating and maintaining knowledge *'gradients'* between the more and the less empowered (Abbott and Jonoski, 2001). It is characteristic of this understanding that it views knowledge exclusively within the ambit of knowledge/power relations. The creation and maintenance of the knowledge gradients has long been achieved not only by acquiring knowledge, which augments power, but also by use of power to canalise, restrict and generally control the flows of knowledge.

In this condition, knowledge has been regarded as a *scarce* resource, requiring a lot of investment. Its production has been either left as a responsibility of the state, or it has been confined within the 'Research and Development' departments of organisations. In both cases, it has been considered as an investment that is necessary, but always with a high uncertainty associated with the expected returns. The drive for efficiency has been understood in such a way that investments in knowledge have to be limited. As a consequence, this drive has been associated with demands for narrow specialisations in different fields, and this has made the sharing and communicating of knowledge even more difficult.

The recent emergence of the field of *knowledge management* needs to be mentioned here because it is closely linked with this understanding of knowledge. It is a relatively new discipline, but when viewed in this postmodern/hypermodern light it is a perfectly understandable development. A system that introduced management of all natural resources, and introduced a 'human resources management' as well, was practically obliged to introduce 'knowledge management'. The main problem of knowledge management is however the fact that in most organisations there is a lack of understanding of the changing nature of knowledge as our societies pass ever deeper into a postmodern/hypermodern condition, even as knowledge in these organisations is approached in the traditional way described above. This is especially the case when organisations with a traditional 'top-down' management structure, whereby the management is largely separated from the principle centres of knowledge production, decide to improve the performance of the organisation through knowledge management. Knowledge is then still considered as predominantly residing in the heads of people (employees), who are often unable (or unwilling) to communicate this knowledge properly. Even as it becomes understood that knowledge is spread out all over the organisation (regardless of its frequently forced 'exile' into certain R&D department) the tasks of knowledge management are still defined as being those of: 'beneficially and more efficiently exploiting the existing knowledge' in an organisation. This is usually understood as creating centralised 'knowledge repositories' which are often extensions of already existing information and database systems, as places for gathering and accessing representations of knowledge, most commonly in textual forms. The stress is here on avoiding redundancies and establishing a certain 'structure' where the existing knowledge can be properly managed through manipulating its representations, and thereby more efficiently exploited. Most of the efforts are then put into a proper abstraction and encoding of knowledge for the purposes of knowledge access and transfer. This approach still puts an emphasis on 'economising' in knowledge resources, and balancing the expenditures on knowledge to a level that will maintain certain 'optimal' *differences* in knowledge, implicitly within the organisation itself and explicitly between the given organisation and both its customers and its competitors. This approach again values knowledge through the *gradient* between the 'knowers' (organisations or individual persons), and those to whom the knowledge is to be provided (either directly, or as an integral part of a differentiated product).

These two aspects of the (initially less postmodern and more hypermodern) changes in attitudes towards knowledge require completely different approaches. The first aspect of knowledge, whereby it becomes a 'commodity', divides the former 'knowers' increasingly into knowledge *consumers* and knowledge *producers*. This shift is so important that it also is sometimes identified as the cornerstone of our postmodern condition (Lyotard, 1979/1984; Appiganessi and Garratt, 1995). The 'commodification' of knowledge in this way is a new phenomenon, but it is already recognised as probably the most powerful in its potential for the transformation of human society. For this reason, it is judged differently by different interest groups in society, and these judgements are sometimes radically opposed. From one side, as we have already indicated, it is considered as a process which leads to a reinforcement of the modern condition and is considered as the primary vehicle that takes us further into hypermodernity. Conversion of knowledge into a commodity is seen by its critics as leading to a further *dehumanisation* of our societies. On a personal, individual, level the process appears so discomforting because it touches upon our capacity for knowledge, considered as exclusive to the human mind, and it is certainly this that distinguishes us as humans. To propose that this precious and indeed for the most part inexplicable 'content' of the human mind that we call knowledge is to be treated as a commodity, to be produced, marketed and sold for profit, is intrinsically disturbing. A large part of this content is still seen as non-externalisable, private, and certainly 'not for sale'. On a social level its appearance marks a new stage in the penetration of money-value relations into the domain of knowledge, and in particular the transformation of knowledge into a commodity means that the value of knowledge is nowadays increasingly determined by the market. The determining factor for any kind of knowledge becomes its *exchange value*. As a consequence, the processes of knowledge production and consumption and the whole 'knowledge industry' are predominantly concentrated in those parts of the world where the market for their knowledge products is most developed and both the supply and the demand for these products is very high. Because these conditions are by far the more predominant in the financially richest societies of the world, the knowledge industry remains primarily their own dominion.

There are indeed a number of reasons for disquiet about this whole process, and some of them, as we shall see later, need to be addressed urgently. On the other hand the significance of the changes and the new opportunities brought about by this process need to be assessed, and transformed into quite other kinds of 'assets' again. The waves of social changes that are generated by knowledge commodification are so strong that they will otherwise wash away any tendency for resistance. The immense technological advances that nowadays enable knowledge encapsulation and transmission have opened the ways to access knowledge that are historically unprecedented. Consuming knowledge is nowadays easier than ever before. Indeed, this ease in the ability to consume knowledge has even changed the perception of knowledge itself. Knowledge is no longer perceived as a *scarce* resource. Knowledge consumerism, to the contrary, shows that, indeed, one and the same repository of knowledge can be sold, exchanged and renewed indefinitely, *without depleting the original store of knowledge!* Knowledge in this sense suddenly appears as an undepletable resource. (Abbott and Jonoski, 2001). (What is perhaps even more discomforting is the fact that our society had to be brought to this stage, where knowledge had already been transformed into a commodity, in order to wake us up to the fact that knowledge is probably the only resource that cannot be depleted.)

This realisation was only made possible through the spectacular technological developments through which the commodification of knowledge was enabled, and knowledge production and consumption were made possible on a *massive scale*. The number of knowledge consumers has grown exponentially, and as a result the unit costs of knowledge production, reproduction, encapsulation and transmission have been reduced by several orders of magnitude. With the processes of knowledge exchange increasing, a new kind of consciousness about knowledge has gradually been achieved, and one which is in sharp contrast to the traditional one. Instead of creating and maintaining knowledge gradients, it becomes more and more clear that what is needed nowadays is to increase the *rate of circulation* of knowledge. Moreover it is being realised that the abandoning of the imperative of a unidirectional flow of knowledge (from the empowered knowers to the less empowered consumers) can actually be more beneficial to the increase of the total rate of knowledge circulation. The key to this realisation is the understanding of the *diversity* of knowledge *needs*. In many different settings, from relations between different organisations, relations between a given organisation and its 'customers' and relations between people in general, it is being increasingly understood – what has always been implicitly understood in person-to person relations - that *different kinds of knowledge can be exchanged for mutual benefits*.

In fact it becomes increasingly clear that individuals, organisations, different communities and social groups value knowledge differently, exactly because of their differences in other areas, such as in their cultural, economic, political or other social conditions. These differences are 'gradients', however not in knowledge but in *values*. The 'willingness to pay' for knowledge comes from these *gradients in value*. Especially after the *awareness* of the existence of these gradients in value is raised, the realisation follows that a particular body of knowledge may be used to sustain, support and uphold certain values. Values on the other hand are as diverse on a personal, individual, level as they are on the collective or social level. And it is precisely this kind of diversity that creates and increases the flow of knowledge.

Whenever the questions of value and valuation are raised, and nowadays increasingly because of the massive scale of knowledge/value transactions, there is a tendency to reduce and transform *all* the *diverse* values into monetary terms. Indeed, many social, and even many personal values can be transformed into money terms, and this is what makes the 'knowledge consumerism' possible. On the other hand, mainly on an individual level (although often on a social level as well), the valuation of knowledge depends on so-called *intrinsic, or inherent values*, which are hardly ever reducible to money terms, but which are nevertheless crucial in determining the flows of knowledge. These values are frequently distinguished from the values that are one or the other way reducible to money terms, by being described as *non-instrumental*. Money as such is never used for anything else but as an *instrument* to achieve this or that goal. In other words, the non-instrumental values cannot enter directly into the processes of *value exchanges* in the way in which the instrumental values can.

The crucial point in these knowledge/value relations is their mutual interdependence. In as much as the gradients in value can increase the flows of knowledge, the improved access to knowledge raises an awareness about the different values and especially about the distinction between the instrumental and the intrinsic values. Although it may appear at first glance paradoxical, the fact is that by allowing the penetration of exchange values in the domain of knowledge, and by thereby transforming knowledge into a commodity that can be easily produced and consumed, a new possibility arises for *removing* the exchange value relations from domains where they have been wrongly, because wrong – mindedly, introduced. This realisation has the potential to work against and even to reverse the penetration of utilitarian economic thinking in virtually all realms of human life.

Since this is a quite complex situation, an example is needed for clarification. In the debate about environmental protection the question of the *value of the environment* repeatedly arises. In any number of projects attempts have been made to assess the 'true' value to this or that environmental feature and then to use this assessment in one or the other procedure for determining 'the preferable course of action'. This procedure is almost exclusively reduced to some sort of optimisation, involving various kinds of trade-offs. It is usually carried out relatively easily once the assignment of values has been made. The problem however resides exactly in this process of value assignment. How do we assign 'the value of a river valley', for example, if there is a question of whether to construct a new dam or not? In reality it is virtually impossible to do this. Assuming that the resettlement of people that have to leave their homes can be assessed through appropriate compensations (and this is quite an assumption already!), what happens with the value of the biological features, the plants and the animals whose habitat will be destroyed? How is one to quantify the values of the specific geological or landscape features? Usually if any of these features is *scarce*, (rare species, unique geological features, etc) the value becomes implicitly higher. In the value assignment process, people (most often 'experts') are being asked to provide *their* assessment of these values. What then happens if there is a feature in the valley that has a unique spiritual or religious importance? Who then takes the responsibility for its valuation? And what happens in the rather ultimate case when people try to preserve a certain area solely for its *existence*, even if they are well aware that they might never go to that area and see it?

Yet, in all projects of this kind the 'valuation' is actually carried out. A so-called 'rational-economic' thinking requires that all these features are somehow reduced into a form with which some sort of cost-benefit analysis can be performed. Methodologies have even been worked out where *existence values* are indeed assigned, and included in the calculations, even while it is obvious that there is no way that this can be done in any 'rational-economic' manner. (For an excellent critique of existence values, see Nelson, 1997). The usual approach of interviewing people and experts has already been proven to be utterly inadequate, usually providing only extremely disproportionate responses. In fact it is becoming obvious that these problems, which usually go under the heading of 'environmental impact assessment studies', *cannot* be solved by this approach. Moreover, the extreme and in fact unresolvable difficulties with valuation frequently open so much space for manipulation by special interests that these studies commonly conclude only with recommendations supporting the most financially powerful and politically dominant interests involved in the process. They even provide these interests with a kind of a *carte blanche* to carry out the projects along lines that are essentially pre-defined long before the assessment has even started. Further to this again, special interests are frequently associated with specific political interests, as exemplified by the sugar-growers associations in India (Roy, 1999), whereby the issues concerned become 'politicised' to a point where even the pretence of 'rational-economic' thinking becomes unsustainable.

In general, most processes of decision making and impact assessment, and especially in the fields of water and the environment, are suffering from similar drawbacks. It is becoming increasingly obvious that problems of this kind must be approached in a completely different way. And the way forward is only through access to knowledge about all the features as well as all the interest involved in the process. This means that rational argumentation, empirical observation, aesthetic pleasure, historical context, traditional experience, moral judgement, testimonies of faith, as much as they are all different and even opposed, will have to be brought together. The processes of 'impact assessment' have increasingly to proceed predominantly through collaborating and negotiating processes that must go beyond the naïve logic of 'rational-economic' thinking. The awareness of the existence of all these different kinds of 'knowledge*s*' is already present. Disregarding this new postmodern condition can only be achieved by some sort of coercion, even if this is applied in highly sophisticated forms.

This situation can only be realised through promoting *learning* processes that are not restricted to isolated and apparently quantifiable aspects of the problems in question. Learning about the state of the problem, the possible solutions and accompanying consequences, must be extended with learning about the values of other persons and interest groups as well as about one's own values and personal interests, described as mentioned earlier by existential writers, such as Marcel, as 'reflective' or 'second order' thinking (Abbott an Jonoski, 1998). It is essentially only through these learning processes that a practice of holistic and integrative thinking can *emerge*, but then with such a critical faculty that these holistic attitudes cannot degenerate into further patterns of dominance and even potentially oppression.

It is in this light that the *opportunities* of knowledge commodification have to be seen (Abbott, 1998; Woods, 1993). The massive learning processes involved can only be realised through a correspondingly massive consumption of knowledge. In this sense, knowledge commodification is compatible with the existence of 'knowledge*s*'. Indeed it may be seen already that there is hardly any knowledge domain that fails to see and utilise as much as possible the opportunities of fast, cheap and easy access to knowledge provided by the new technologies that support the knowledge industry. The objection frequently raised is that the knowledge industries are motivated only by increases in profits, and that they are in fact very far from being directed towards the aims that are presented here. Even as we accept this, we have still to insist that there are great opportunities inherent in this process. Knowledge products, like all other products, need customers. Successful industries have a continuously increasing customer base. One of the greatest potentials for this increase lies exactly in the processes described above. The readiness of people in any community to 'buy' knowledge in order to achieve a harmonisation of their individual and collective intrinsic values, with the social, including the instrumental values emerging and establishing themselves within that community, is indeed already beginning to appear within certain isolated contexts. (Community is here used in the broadest possible sense and it refers to any kind of social organisation). This harmonisation is the key factor for achieving sustainability of any community (Abbott, 2000a). In some sense it is the same struggle that has been going on throughout human history: trying to provide a decent environment where the social and the individual can live at peace. All the battles (sometimes indeed horrifying and bloody) between the two have been because each was blind to the values and importance of the other.

This new nature of knowledge, as we have seen, causes our attention to turn away from the exclusive study of knowledge/power relations and to direct our gaze more towards knowledge/value relations. In fact, in any discourse on knowledge that focuses uniquely on knowledge/power relations, the *dominant* position of the knowledgeable and powerful appears as inevitable and hardly questionable. The shift towards the knowledge/value relations, on the other hand opens up the possibility to conceive of *diversity without domination*. At this moment it may indeed appear only as a possibility. But by building an understanding of the complex of knowledge/value relations and by using the opportunities opened up by the commodification of knowledge, this shift can become the key factor for sustaining the future developments that alone can lead towards knowledge consumption on a massive scale for more positive purposes.

At the same time, *mass consumption of knowledge*, which can indeed be the most promising development for the future of our societies, is still only a possibility and it will take many more and more various efforts to achieve the desired effects. The problems ahead are then of course far from being only technical. In fact, the problems related to technology will almost surely be the more easily addressed. One aspect that becomes more and more clear is that the achievement of mass consumption of knowledge cannot be achieved by the traditional methods used for achieving the mass consumption of other products. The tendency to shape *uniform* needs in order to satisfy the requirements of mass production is certainly not going to bring the desired outcomes. The diversity of knowledge needs is genuine and legitimate and must be addressed as such. This becomes obvious as the diversity of demands for knowledge bring with them the imperative of providing customised and even personalised knowledge products, and especially in the case of those products that are provided over electronic networks. In many cases this appears only as a question concerning the form of delivery, but there are already more and more products which provide customised content in daily uses and then provided especially by the financial sector of the economy. Through these processes, the mass production and consumption of knowledge can in fact gradually lead towards a *demassification* of the culture, including its institutions, and then precisely in the societies where it is mostly developed (Dyson et. al, 1994).

However, the greatest challenge facing future producers and distributors of knowledge products is the second aspect mentioned above, namely the mobilising of the various kinds of knowledge in such an integrated way that the harmonisation of individual and the social values can be realised. This task necessitates a fostering of the processes of social and individual learning, knowledge sharing and collaboration. The very nature of these processes indicates that, for this purpose, attention has to be shifted away from the body of knowledge itself, and directed more towards providing *environments* or *contexts* suitable for knowledge circulation. Future knowledge products, even though they may carry this label of 'knowledge', will not be directly concerned with knowledge itself, but more with providing appropriate environments where the processes of knowledge production and exchange will be facilitated. It is interesting to notice that even at the level of corporate knowledge management the focus is beginning to shift away from structuring existing repositories of knowledge and towards providing proper environments that will accommodate the processes of knowledge sharing, and so stimulate human innovation (see for example Gergen and Thatchenkery, 1997; McElroy, 2000).

Building such environments is precisely a sociotechnical task. The next generation of technical tools will less resemble systems where representations of knowledge are just processed through input-output relations, but will be much more environments for better enabling, and thereby enhancing, the processes of knowledge circulation. They will have to utilise the present and future electronic networks as platforms upon which they can be constructed and deployed, simply in order to facilitate the establishment of networks of people and organisations. Many of the key processes in the future networked societies, including the processes of decision making, will be potentially carried out through the utilisation of these sociotechnical networks.

However, the beneficial use of the new condition of knowledge through sociotechnical design will still have to confront many new challenges. Technological advances that are enabling the processes of knowledge circulation can be, and in many cases are, equally utilised for maintaining and even increasing the differences between the more powerful knowers and the less empowered consumers of knowledge. The inequality in access to knowledge is increasingly measured by inequalities in access to electronic networks such as the Internet. In fact the traditional division of the three worlds, with its primarily ideological and geographical connotations, is now transforming under the influence of the introduction of electronic networks of knowledge and data. The first world consists of those who have access both to the 'content' (the electronically encapsulated reserves of knowledge) *and* the possibility to influence this same 'content'. The second world consists of those who only have access to the body of knowledge but do not have the possibility to influence its content. The third world consists of those who neither have access nor can influence the 'content' so that they are often treated by the first two worlds as 'mere objects' (Abbott and Jonoski, 1998). This division is still however an ideological one and the ideology behind it is itself one of the first world. Therefore the possibility for abusing and terrorising the third world by the first and the second one is still very much present. Knowledge degenerates into a particular obnoxious form of violence when so employed against the 'others' (humans, our fellow creatures and the natural environment itself).

It is essentially for these reasons that the debate about the so-called 'digital divide', or 'electronic divide', gains so much attention at present. It is becoming increasingly clear that the primary reason for the large inequality in our human societies, both on a global and on local scales, is an inequality in access to knowledge. At the same time, however, it seems that the 'mere objects' from the third world have little to hope for. The habits of using knowledge merely for exploiting, and in some cases even terrorising, the third world are hardly changing, even as the rhetoric of social change that is being introduced through the processes of knowledge commodification and dissemination seems to be spreading everywhere.

Thesis 6: *Simulacra provide not only the means to increase the circulation of knowledge by some orders of magnitude but also the means to bring persons of very diverse backgrounds together around representations of actual realities. The Internet–enabled computer game, which allows for multi-person participation and frequent real-world refreshment of its realities appears as particularly well suited for promoting second-order reflective thinking among its players.*

2.7 The so-called end of history and the third world

In the summer of 1989, an article entitled "The End of History?" appeared in the American journal *The National Interest*. In the dusk of communism in Central and Eastern Europe, its author, Francis Fukuyama, put forward the argument that the world had indeed reached the 'end of history'. The liberal state combined with the free-market economy was proclaimed not only the victor against communism, but also the culmination of all human historical development. At the time, the article sparked unprecedented intellectual reactions, which were far from univocal. A year later, the author came up with a whole book entitled "The End of History and the Last Man" (Fukuyama, 1992). The Central and Eastern European 'velvet revolutions' that had occurred in the meantime were probably encouraging for dropping the question mark and adding the 'Last Man' part. In sharp contrast with the postmodern spirit, which reveals that there are no universally applicable rules that will save the planet and bring a new humane order on Earth, Fukuyama proclaimed (1989):

> What we are witnessing is not just the end of the Cold War, or a passing of a particular period of postwar history, but the end of history as such: that is, the end point of mankind's ideological evolution and the universalisation of Western liberal democracy as the final form of human government...

In his book the author tries to present the picture of a completely rationalised historical process that is moving upward (by way of 'progress') and how the recent events inevitably lead us to conclude that we have finally reached the summit of human endeavour. In a typically modern fashion, the line of historical development is traced from Plato to the present, while relying heavily on Hegel (and one particular interpretation of Hegel, of Kojève), in order to persuade the readers that the ideological development of humanity, after a great struggle, has finally attained its peak, and that we are about to enter a 'post-historic' era. The 'Last Man' is the one who achieves this sublime state of fulfilment. The 'perfected' human nature which was 'yet to come' and was 'unrealised', or present 'only in potential', is now sought through the realisation of this new variation of Hegelian Grand Narrative. Fukuyama admits that human life in this post-historic era might be 'dull', and sometimes even 'very sad', because the things that once called for 'daring, courage, imagination and idealism' would now be replaced by economic calculation. Nevertheless he maintains the essential Hegelian imperative, that this is a *necessary* development, that cannot be stopped by individual contingencies. Our purpose here, however, cannot be to analyse the book itself. There were plenty of reactions to his thesis (for one of the more illuminating see Kimball, 1992). It suffices to say that, although it may appear to some as eruditely written, Fukuyama's philosophical account is in fact not at all impressive and even less persuading.

What is much more important is the question of why such a book became so popular and triggered such a strong intellectual response? Was it just because of an 'end of millennium' spirit, obsessed with 'endisms' of all possible kinds, and for any number of different reasons (end of history, end of ideology, end of Man, end of art, end of philosophy, etc.)? Or was there another, deeper reason for re-stating the necessity and inevitability of an all-winning Grand Narrative at the moment when all threats and alternatives seemed to be defeated.

The jubilation of the 'end of history' was particularly loud throughout the first world establishment. In fact it was found to be extremely useful as a sort of 'theoretical' support for the *current* policies in the first world that had already been on their way during the 1980s. The shift towards 'deregulation', 'market liberalisation', and the reduced role of the welfare state were already establishing themselves as key steps towards future *economic growth*, which was proclaimed as the only possible path towards prosperity. The 'end of history' came to be interpreted for all practical purposes as identical to the so called TINA ('There Is No Alternative') slogan which became a sort of New Gospel during the last two decades of the 20th century. The linkage between these two slogans actually created a powerful argument in support of the *universal applicability* of these policies, with the frequently repeated promise that they were guarantees for prosperity even for the third world societies.

The most vocal warnings that this was a completely false expectation, or optimism, created and supported by neo-liberal ideology, came from postmodern thinkers. Derrida warned that this kind of celebration of liberal democratic capitalism in facts serves only to hide how much this 'final state' is threatened and is itself threatening. In his words (in Appignanesi and Garratt 1995, p.170):

> ...never have violence, inequality, exclusion, famine, and thus economic oppression affected as many human beings in the history of the Earth and of humanity...no degree of progress allows one to ignore that never before, in absolute figures, have so many men, women and children been subjugated, starved or exterminated on the Earth.

Indeed, a decade later, in a world dominated by global capitalism, there are hardly any signs that the third world societies will see the neo-liberal promises materialising in any near future. The World Development Report 2000/2001 issued by World Bank (2001), concentrating on poverty alleviation, presents clearly the grim picture:

> The world has deep poverty amid plenty. Of the world's 6 billion people, 2.8 billion - almost half - live on less than $2 a day, and 1.2 billion - a fifth - live on less than $1 a day ...
>
> This destitution persists even though human conditions have improved more in the past century than in the rest of history - global wealth, global connections, and technological capabilities have never been greater. But the distribution of these global gains is extraordinarily unequal. The average income in the richest 20 countries is 37 times the average in the poorest 20 - a gap that has doubled in the past 40 years.

A similar picture and even a similar wording is presented in reports dated 10 or 15 years ago. In spite of various programs for development and development aid, little has been achieved to improve this situation. To the contrary, all data suggest that the gap between the rich, primarily in the first world, and the poor, primarily in the third world, is only increasing. Indeed, so far has this situation deteriorated that even such euphemisms as 'developing countries' have become discredited.

It is indeed mainly in the countries that traditionally belong to the so-called third world that the problems of poverty and underdevelopment are most severe. In these countries the day-to day problems are far from those existing in the first world. The objectives of many people living in these countries can hardly go beyond the demands for bare survival. However the patterns of social exclusion are existent in many of the more developed and even in the first world countries. These are what Castells (1997) calls the 'black holes of informational capitalism', that are not confined to any particular territory. In the light of the disappearance of the statist second world and the lost significance of the third world as a relevant entity, he names this new, heterogeneous, world of poverty - the fourth world (Castells, 1998, p164):

> The Fourth World comprises large areas of the globe, such as much of Sub-Saharan Africa, and impoverished rural areas of Latin America and Asia. But it is also present in literally every country, and every city, in this new geography of social exclusion. It is formed of American inner-city ghettos, Spanish enclaves of mass youth unemployment, French banlieues warehousing North Africans, Japanese Yoseba quarters, and Asian mega-cities' shanty towns. And it is populated by millions of homeless, incarcerated, prostituted, criminalised, brutalised, stigmatised, sick and illiterate persons. They are the majority in some areas, the minority in others, and a tiny minority in a few privileged contexts.

The number of these people belonging to the 'fourth world' is continuously increasing. It is estimated that with the continuation of the current trends, by the year 2025, there will be around 2 billion adults at their working age who will hardly posses even the basic skills for any other job except the most basic manual labour. With the parallel processes of globalisation of illegal activities through increasingly connected criminal networks (see again, Castells, 1997, 1998) these people will become easily recruited into this criminal economy, as this becomes their only exit from life in destitution. This can only increase the pressures and bring more tension in all human societies, regardless of whether they are of the first, second, third or fourth world.

In the light of this reality, the discussions around the 'end of history' and proclamation of 'victories' seem already obsolete and irrelevant, and they are indeed now already largely forgotten. Yet, the same spirit still dominates much of the current supporters of the processes of globalisation and, in particular, it is this that directs the mainstream approaches to problems of development. Focusing on increasing economic performance only (increasing GDP being then the guideline for all development investments), has been seen for a long time in terms of primary objective of stimulating the *physical* elements of economic production. Developing systems and activities that focus on development of *people and institutions* have long been largely neglected (Woods, 1993).

Coupled with the requests for market liberalisation, deregulation and restrictions of government spending, this approach has in fact proven to be detrimental to the development of what is called *human* capital, which is seen by the critics of the current approach as far more important for overall development. The 'strings' that are inevitably attached to large development funds (mainly loans provided by International Monetary Fond –IMF- and the World Bank) have frequently contributed to reductions in government spending in crucial sectors such as health and education. Terms such as 'over indebtment' and 'the debt trap' have become common currency. This has led to the formulation of so-called Structural Adjustment Programs – SAPs- which however have already proven to be not only inefficient in achieving their development goals, but in fact destructive for whatever limited potential that still existed in many third world societies for indigenous development (see for example Haywood, 1995). The domain of 'development' is probably the most misplaced precisely in these places where such socially destructive policies have been introduced, and all in the name of *efficiency* and *performativity*.

The whole approach taken by the most directly responsible agencies involved in development has come recently under severe criticism. The Multilateral Development Banks (MDBs), with the leading role of the World Bank, and the IMF, have been accused not only of applying ill–founded development programs due to a lack of understanding of the real problems in the third world societies, but also for their lack of *will* and true *commitment* to address these problems. Many critics often see them only as proxies for supporting the interests of the large multinational corporations and of global capital generally. Instead of stimulating economic growth in the third world that would alleviate poverty, the current development model as often as not achieves the opposite. It protects the interests of the governing elite in already economically-stratified third world societies, it promotes the interests of the transnational investors and contractors and it supports the overall globalisation agenda of the donor governments of the rich first world (Karliner, 1997).

Regardless of the variety of development policies, most of them fail to achieve the officially stated goals, and their final outcome contributes to the picture presented by their critics. One of these policies is the continued funding of large infra-structural projects in various sectors such as water, energy, agriculture, forestry and transport. Frequently the social and environmental costs associated with these projects are not taken into account and emerge only after finalising these projects. The situation in the water sector particularly exemplifies this situation (Abbott, 2001). The impacts of large water engineering projects, especially the construction of large dams, has recently caught the attention of a very large audience that exceeds the traditional circle of water specialists. (See for example the "Dams and Development" report (2000), issued by the World Commission on Dams, and all the reactions to it, as one of the results of this broadening awareness).

In fact, it is becoming increasingly recognised that the linkages between the natural environment and the social structures in the third world societies are far more complex than previously realised. The consequences of environmental destruction for these societies are frequently bringing into question the very survival of the affected communities. The Sardar Sarowar dam project on the Narmada River in India has become a widely known example. It even gained symbolic significance because of the successful public campaign against forced resettlement of people, that eventually forced even the World Bank to cancel its funding for the project (For this remarkable struggle of the affected people, see Roy, 1999). A Canadian environmentalist tried to explain this fight in words closer to her first-world audience (Adams, 1994):

> At issue in the Narmada River valley is land. The people just do not want to move. They want their property rights to their land, their water, their forests – in short their ecosystems – respected...Their expertise: what to plant, how to cultivate, and when

to harvest, is completely bound up in a particular ecology and determined by the seasons, the soil, and the vegetation they know. Their vast legacy of unrecorded knowledge about their environment, and how to survive in it, becomes almost useless in another.

That is why eminent American anthropologist Thayer Scudder, who has studied the effects of forced resettlements in many Third World countries says, "It is about the worst thing you can do to a people – to force them to move - next to killing them. That may seem overly dramatic to us. While we, well off Americans and Canadians can imagine the anger and the sorrow of having our homes expropriated and our communities destroyed, few of us would take it to the point of suicide. But for many people in the Third World, expropriation and destruction of their bioregion is a matter of life and death. Their land is their means of subsistence. Taking their land from them is the equivalent of taking from us not only our land and our homes but also our bank accounts, our investments, our education, our professions, and our social security systems.[8]

Virtually all the policies of the prevalent development model are coming under attack by some hundreds of smaller organisations, such as many Non Governmental Organisations (NGOs) of various kinds, environmental and labour groups, and organisation representing the interests of local communities. The obvious failure of this model has brought about the discussion of the need for changes even in the first world establishment (see Abbott, 2001). What is of particular interest here however is the way in which such large number of different organisations have managed to mobilise new social forces, sometimes referred to as the anti-globalisation movement. The use of Internet and all the advantages that are offered by the new communication technologies have been crucial in their actions. Organised through hundreds of web-sites, this movement is showing how different 'kinds' of knowledge and information can be efficiently communicated, and that these simply cannot be streamlined or controlled by any 'establishment'. There is hardly any structure in the organisation of all these different groups and organisations, yet the shear speed of information exchange powered by modern communication technologies enables them to create and mobilise substantial physical forces. And even though most of their activities are being reported in the mainstream media only if they cause disruptions in parts of the first

[8] The 'pro-development' circles blame exactly these attitudes and similar opinions, for in fact sentencing the people in the third world societies to a continuous life of underdevelopment. They claim that without openness to the world economies the third world societies will be never given the opportunities to effect positive changes in their lives. Open markets and free trade agreements are proposed as in fact beneficial for all. The crucial argument that is often presented is that these policies provide the 'levelled playing field'. The problem with this approach is clearly expressed by Castells (1996, pp.87):

> Naturally, since "free and fair market conditions" belong to the unreal world, political agencies acting in the international economy seek to interpret such a principle in a way that maximises the competitive advantage of firms under their jurisdiction…

Moreover, even if one can imagine a 'levelled playing field', without the necessary human capital and institutional strength third world societies are predestined to be losers in the new game of global economic competition. There may be a levelled playing field, but one team has the world class players, while the other is in poor health and largely untrained.

world, a more thorough search through their various information sites shows that their action is also truly global.

This example inevitably takes as to the key question about the potential and the role of the new knowledge economy - which now becomes so dominant and determining for the societies in the first world, - in relation to third world problems. At present it certainly looks as if the gap in wealth between the two worlds is only reinforced by their gap in knowledge. The commodification of knowledge currently shows no signs of spreading rapidly into third world societies. There are two processes, however, running in parallel, that are compatible with the two different understandings of knowledge presented in our previous section. Their mutual relationship will probably determine whether the knowledge economy will bring benefits to the third world, or instead will bring only further deprivation. From one side the traditional understanding of knowledge as 'power' will further increase those processes in the knowledge industry that try to maintain the knowledge 'gradients'. In the open knowledge industry, this means that more and more knowledge products, and especially those that are perceived as crucial for maintaining 'competitive advantage' will be protected by copyrighting and patenting. In the context of relations with the third world, this means that the traditional 'knowledge goes south, things come north' scenario will continue, and will become even more exacerbated. The expertise coming from the first world will not be limited to technical or financial assistance, or even to direct investments in third world economies. The deteriorating institutional structure in the third world societies will hardly be able to sustain even the basic needs of its people. This can only lead to a situation where domains like health, education, and social services will become locked in their total dependence on the knowledge provided by the first world, which will only be delivered at a price. When seen from that side, the commodification of knowledge seems to offer little hope for the third world, and in fact even appears to worsen the situation. As Haywood explains (1997, p.125):

> One clear concomitant of this scenario is that more and more knowledge domains, and the informational infrastructure that supports the maintenance of tradable stores of knowledge, would have to be held within secure, excludable, 'commercial and in confidence' boundaries. Knowledge that was once shared on a fraternal basis could be jeopardised by such a 'trade only' policy. In reality it would, of course, be difficult to know when or how to switch information from a fraternal to a 'trade only' status. However, the possibility that developed countries could begin restricting the kind of information flows that once were exchanged freely, out of deference to new business imperatives that sought to recatalogue the commercial value of the exchange, is one that could strike a mortal blow to the embryonic research and development programmes of many developing nations.

An additional danger to this 'restriction of information flows' is the penetration of the attitude that 'anything goes' into the domains of knowledge, which can be extremely successful in business terms when dealing with unskilled third world customers with a low capacity for judging the value of the knowledge products delivered to them. It is clear that this attitude, when it comes to knowledge, will have disastrous consequences, but experience with the more traditional 'transfers of expertise', where all kinds of obsolete technologies were dumped in third world societies, shows clearly the potential dangers.

The second process is the one closely related to the 'knowledge circulation' approach described in the previous section. The tendency to protect the knowledge 'surplus' in the first world has to be contrasted with the potentials for opening new markets for the knowledge products themselves. The third world is certainly poor, and it is impossible, outside of one or two very special localities, to expect such large financial returns on the investments in its knowledge industry. Yet, the needs for knowledge there are far greater than in the first world, especially when knowledge is the only 'weapon' that the people from the third world can use in their struggle for survival. The kinds of knowledge products needed in the third world are, however, different from those of the first world. The population will only accept these products if they prove to be helpful in resolving their immediate, day-to-day practical problems. These products must go beyond the mere exhortations concerning the need for environmental preservation and sustainable development or proper management of their resources. As Woods explains (1993):

> Educating people *about* natural preservation is not enough. The education and training to achieve that preservation is needed. Interactive technology has immense potential as a tool for assisting communities to enter their own data on their own resources and to process it and weigh up alternative solutions. It carries the learning process into practical application in specific situations and can then provide technical instructional materials to suit the conditions of these situations. This capacity to help individuals, groups and communities to manage their own resource is being shown to have a remarkable 'mobilising' effect.

In addition to these aims, knowledge products will have to provide *advice* on such various issues as potentials for marketing of locally produced goods and access to broader markets. The only possible for fighting poverty is providing people with the means to pull *themselves* out of poverty. This is clearly exemplified by one of the most successful projects that has been carried out in rural Bangladesh - the microbanking activities of the Grameen bank, which is now being replicated in many other countries (Yunus, 1993//1997; 1997). There, the actual credits provided to the poor are linked with appropriate advice for all kinds of possibilities for economic development. Sectors as diverse as agriculture, fisheries, textile manufacture, energy and telecommunications are now part of the Grameen family. Through a commitment to work directly with the people who are in real need of assistance, Grameen bank has established a genuine grassroots-driven development.

These kinds of developments offer the real opportunities for future knowledge products in the third world. It is only through a linking with such genuine grassroots movements that the new knowledge products can bring the expected benefits (see again Abbott 2001; Abbott and Jonoski, 2001). At the same time, this process needs the mobilisation of all the existing local potentials (sometimes referred to as 'knowledge islands') that can contribute to the design and development of the new kinds of knowledge products. These local centres will have to cease serving only as extended arms for channelling traditional flows of development funds, or as sources of cheap labour for software development. Instead, they will have to start looking upon their own third world as a potential market.

It will nevertheless be illusionary to think that the knowledge industry will be able to bring benefits to the third world without a joint action by governments, the business sector and especially the large lending agencies. The knowledge industry cannot work without a proper communication infrastructure, and this requires a considerable investment. Traditionally, the lending agencies did not invest very much in this sector under the assumption that this could be left to private sector investment. These attitudes are beginning to change, but an immense work is still required for bringing the basic communication networks up to an adequate capacity, albeit still far from those already operating in the first world. While the role of the lending agencies in this task of providing the basic infrastructure is indispensable, it is even more important for them to change their attitudes and even completely redefine the notion of 'development'. The recent transformations in the World Bank, the leading lending agency, and the one that sets the example for most of the others, show that attitudes are in fact beginning to change. The World Development Report from 1998/1999 for example (World Bank, 1999) clearly identified the crucial role of knowledge in development. The Bank's commitment to its transformation into a 'Knowledge Bank', is another example that this leading institution is taking the role of knowledge seriously. By establishing several gigantic web-sites (the costs of one of them, the so-called Development Gateway, are estimated at 60 million $US) for sharing knowledge on issues related to development, this institution may take the true leading role in dealing with development problems. There are signs however, that even these activities are again only concealing the traditional policies behind the new rhetoric. Careful analyses of the developments so-far suggest that the embracing of the Internet by the World Bank may be with the intention of monopolising the complete discourse on development and keeping the development agenda under its firm control. In a thorough analysis of these recent changes, King and McGrath (2000) ask several key questions, for which there are still no clear answers at the present stage of the World Bank transformation. One of these key questions remains the following (ibid):

> In the 1980 and 1990s the Bank's strong identification with the Neo-liberal orthodoxy meant that it was perceived to be very much in the business of telling others rather than listening to others. In stressing the new identity as a knowledge bank, has the organisation been able to move away from that view of being a knowledge provider to one in which it is also a knowledge seeker and sharer?

Many organisations are already protesting against the ways in which the World Bank seeks to establish itself as the 'Knowledge Bank' on development. Some of them are even calling for boycotting cooperation with the Bank's new web-platforms. Most of these reactions are mainly due to the already badly empaired reputation of the World Bank. It is of the utmost importance, however, to continue the dialogue and the pressure also on this institution, for its own transformation. A World Bank, which is much more transparent and prepared to cooperate with grassroots organisations, would be a valuable adjunct to the future of third world development[9].

[9] Much of the impetus for a reform of such large organisations as the World Bank is generated from the impression that the provision of knowledge within third world societies will be immensely expensive affair. This impression no doubt arises from observations of the massive investments in communication technologies and infrastructures observed in the first world. Moreover it has been the mark of every sociotechnical revolution (canals, railways, steamship services, etc) that these were all hugely profligate with investment. Abbott (private communication) speaks of a 'human wave approach' to challenges in the new communication technologies, whereby hundreds of thousands of people, and countless billions of dollars are thrown at problems in largely uncoordinated and thoughtless manner (see, also Wolff, 1998, and Easterly, 2001). In an article printed in

Finally, at this point, it has to be stressed that even if all the positive developments within the lending agencies come to fruition, and even all practical business opportunities are realised, a large number of individuals will not be able to enjoy the benefits of this new approach at any time soon. There are still many people who lack the basic skills for even acquiring any knowledge, let alone producing and sharing. The interests of many of these 'mere objects' of the third world will remain to be represented by interest groups that are themselves situated, at least in the foreseeable future, predominantly in the first world. It means that these groups will have to make a special use of the new knowledge sharing tools and services, particularly in order to disabuse individuals and organisations of the habit of using knowledge *against* the third world. The hope is that because they will be able to mobilise this knowledge in their actions it will be possible to avoid the 'double victimisation' of the 'third world', once by first-world society and a second time by those who represent them in that same first-world. But, unfortunately we have to realise that "the indignity of speaking for others" (Owens 1985) cannot be avoided completely, here as elsewhere.

Thesis 7: *Knowledge circulation processes in the third world have positive potential for fighting poverty. The enabling of these processes by introducing new kinds of sociotechnical arrangements should be considered central to a new development agenda directed towards promotion of increased mass consumption of knowledge and with the goal of achieving self-reliance of third world societies. Such an agenda should replace the current inefficient and often detrimental development models which largely support the corporate interests of the first world instead of supporting the interests of the poor in the third world.*

2.8 Hydroinformatics as sociotechnology

Although the necessity of taking a sociotechnical approach to the problems that we are facing in our present condition becomes obvious in many different and previously isolated domains, our most immediate concern is with the domain of water and the aquatic environment. The indispensable role of water in both the natural and the human economies has been re-stated any number of times. In our new, networked societies we are facing an ever increasing complexity in our relations with water. Such is this complexity that it requires a new paradigm for dealing with water and environmental problems. This new paradigm is what we here call sociotechnology, and when it comes to water the most appropriate candidate to take this paradigm wholeheartedly is the discipline of *hydroinformatics*. In order to see why hydroinformatics is the discipline most prepared to embrace the new approach we need to turn to the foundations of the subject, and follow the ways in which it has been transformed and developed during its short existence of only one decade. This study is a hydroinformatics

Financial Times, William Easterly said that the $1,000 billion spent on aid to developing countries, with the help of the IMF and World Bank "have all failed to attain the desired results". The approach that is supported by the present study calls instead for a well prepared and properly coordinated approach that will be comparably less expensive, and certainly far cheaper than continuing the drive to mechanisation and the intensified use of agrochemicals which are the only practical alternative.

study, but it is also a study *about* hydroinformatics, and in that sense this reflection is necessary.

Hydroinformatics is *essentially* a technology (Abbott, 1991). It is a technology in Heideggerian sense, which finds the essence of technology in revealing, or unconcealment (Heidegger 1963//1977). Heidegger made the important distinction between technology *in its essence* and *modern* technology. Although he grants a basic correctness to an instrumental view on technology, that it is an assembly of means for achieving ends, he insists that the essence of technology resides in quite another place, namely in revealing and unconcealment in such a way that we experience *alētheia*, or *truth.* The origin of the word *technology* is in the Greek *technē*, which marked the unity of all creative, or *poietic*, human activities. Both the arts and the work of the craftsman therefore belong to the realm of *technē*, and *essentially* they are the same. Technology is thus a human *poietic* action in which revealing takes place "and where alētheia, truth happens". The way in which this revealing takes place is through a process of 'bringing forth', or 'coming to presence', which results from the essential unity of all factors of *causation.*

Even in modern technology the essence remains unchanged. The difference that Heidegger distinguishes in *modern* technology resides in the way in which this 'bringing forth' and 'coming to presence' happens. In modernity, this proceeds through 'challenging', "which puts to nature the unreasonable demand that it supply energy that can be extracted and stored as such" (ibid). The ultimate mode in which this challenging is happening is the transformation of everything (as already mentioned several times in the preceding sections) into a 'standing reserve' (Heidegger uses the word *Gestell* -Enframing) through the processes of 'ordering', 'counting' and 'computing'. This challenge however, becomes the greatest danger, as it transforms humanity itself into 'reserves' and 'resources', and in this process it blocks the essential thrust of human-technological action –its revelation of truth. Heidegger explains that modern technology, through these two processes, harbours in itself both the greatest danger of all, in the sense that it becomes the highest stage of misrepresentation of the essence of being human, while, at the same time, it is inherently the 'saving power', because of its essence, which is its revelation of truth. Thus (ibid, p.33):

> The essence of technology is in a lofty sense ambiguous. Such ambiguity points to the mystery of all revealing, i.e, of truth.
>
> On the one hand, Enframing challenges forth the frenziedness of ordering that blocks every view into the coming-to-pass of revealing and so radically endangers the relation to the essence of truth.
>
> On the other hand, Enframing comes to pass for its part in the granting that lets man endure – as yet unexperienced, but perhaps more experienced in the future – that he may be the one who is needed and used for the safekeeping of the coming to presence of truth. Thus does the arising of the saving power appear.

So when we say that hydroinformatics is essentially a technology, it should be clear that it is not a modern technology in the sense presented above. It is still however to some extent modern technology in the sense that it employs modern physical science in its process of ordering, counting and computing in order to produce 'knowledge reserves' concerning waters, which are in term conceived almost exclusively as "water reserves' or 'water resources' (Abbott 1994). Since Heidegger identifies modern physical science with Enframing ("Modern physics is the herald of Enframing, a herald whose origin is still unknown" (ibid, p.22)), it is clear that hydroinformatics has to cope itself with the ambiguity of technology mentioned above. For hydroinformatics, coping with this ambiguity meant that it must be aware of its *social* dimension from its inception.

From its very origins, hydroinformatics accepted that modern technology is not just the application of modern physical science. Although science frequently contributes to technology, these are only the points of intersection of two lines of thinking that are not even approximately following the same directions. Indeed, technology is more often than not developed under quite other influences then those of science, such as those of aesthetics or business. Or, as Lyotard (1979//1984, p.45) puts it even more pointedly referring to our present condition: "The 'organic' connection between technology and profit preceded its union with science".

The actual birth of hydroinformatics coincided historically in the most basic technical terms with the transition from the third to the fourth generation of modelling. If the first three generations of modelling had been developments in hydroinformatics in depth (being more science driven), the fourth generation was already a development in another direction, this being more in breadth of application, placing much more effort on user interfaces, new types of graphical outputs, parameter controls and structured date-base facilities. These efforts were much more pushed by other influences (and even less science driven) in order to enlarge the number of people who could apply the electronically encapsulated knowledge. With the fourth generation of modelling systems, the number of users to whom the revealing has been offered has already increased to many thousands (see Abbott et al. 2001). These are still predominantly specialists in hydraulics, hydrology and water resources "placed predominantly within the corporate context of a company, a government, a government agency, an investment or insurance organisation or similar body"(Abbott, 1996). Through these developments, hydroinformatics already began to demonstrate its social dimension. Its employment has profoundly influenced the social structures of these corporate bodies, as well as reshaped the relations between these bodies (Abbott 1996).

It was initially through these processes that hydroinformatics began its transformation into a sociotechnical discipline. One of the key characteristics of its sociotechnical character was its rejection of various separations introduced by modernity (see section 2.2). Hydroinformatics recognised the 'hybrids' openly, and indeed employed these in order to expose and even create them. At the same time it can be said that through these processes it begun, also, to influence the separation of knowledge. The electronically encapsulated scientific knowledge incorporated into hydraulic, hydrological, ecological and other kinds of modelling systems has been made available to a great variety of specialists from different fields and these have combined this knowledge with all manner of other kinds of knowledge. It is not only that the boundaries between different fields were thus made more flexible and a new broader discourse was opened up, but also, through the increased number of people that had access to this knowledge, the general awareness of the 'hybrid' nature of problems related to water and the environment was increased.

But how was this achieved? The key transformation that had occurred during the emergence of hydroinformatics from its field of origin – computational hydraulics – can be identified with a shift from a *symbol*-representation into a *sign*-representation of knowledge in hydraulics and hydrology. While symbols, operating as tokens, function by *replacing* features of the real world into the minds of the water engineers - signs work by the virtue of their *pointing towards* these features. The significance of this shift is such that we have to speak of a *paradigm shift*, where the new paradigm is called *post-symbolic*. Only due to such a shift was the design and development of sophisticated modelling systems that encapsulate *generic* knowledge made possible, and by virtue of that process this knowledge was made available to such a large number of people *outside* of the previously small circle of numerical modellers As Abbott et al. explain (2001):

> In principle, a numerical-hydraulic modelling tool is one that encapsulates generic hydraulic and related knowledge in such way that this may operate on site-specific data in order to provide site-specific knowledge. The extent to which this knowledge comes to presence in the mind of the tool user naturally depends upon the ability of that user to interpret the (mostly graphical) output of the system, and so upon that user's particular *knowledge frame* (Abbott, 1993). A considerable part of the effort put into fourth generation modelling systems has been directed to increasing the rate of knowledge processing by their users by the provision of more appropriate interfaces and supporting facilities, while most of the rest of this effort has been expended upon extending the range of application of these systems and automating their instantiations. Thus, whereas in 1985 some eighty percent of investment was still associated with the numerics of modelling tools, by 1998 at least seventy percent was being used for enhancing knowledge transmission facilities. The emphasis has passed correspondingly from number organisation and application, or *numerics,* to sign production and organisation, or *semiotics.* It has thus passed over to the study of sign production, distribution and consumption, which belongs to the subject area of the *theory of semiotics* (e.g. Eco, 1976). In the simplest terms, we observe the transformation:
>
> **Numerics → semiotics**

Hydroinformatics as a whole, and *especially its social dimension,* is primarily associated with this shift. The generic definition of a model, from a hydroinformatics point of view, presents this even more clearly (Abbott, 1992, 1993, 1994): "A model is a collection of indicative signs that serves as an expressive sign". The emphasis here is not on 'a simplified representation of reality' or on 'a representation of systems for converting set of inputs into set of outputs', as in most other common definitions of a model, because these become too restrictive for the purposes of hydroinformatics. The fact that the choice, number and arrangement of the indicative signs in order to produce the expressive signs *determines* the very expression that a certain model creates, puts the whole emphasis on the way in which models serve as devices for *communicating* knowledge. And this way of knowledge communication in all our modelling systems is through the creation of expressions, in other words semiotic devices that induce experiences, including experiences of *phenomena.* It is then through the employment of appropriate *sign vehicles* that a phenomenological approach is made possible which offers much more flexible and more *persuasive* knowledge communication, particularly suited to the increasing diversity of the knowledge *needs* of the users of the hydroinformatics systems. It is well known that within the previous symbolic era, the emphasis on knowledge *representation* itself introduced limitations on knowledge communication, and the processes of knowledge production and dissemination were therefore limited to a much smaller group of people.

This of course does not mean that modellers have the freedom to choose arbitrarily the type, number, and the arrangement of the indicative signs used in their model building. Models still have to be built to provide as accurate representations of physical realities as possible, but the ways in - and purposes for - which this representation can be communicated to the various users are now immensely increased. In order to further deepen the understanding of this process, the design and development of future hydroinformatics systems have to rely heavily on the field of *semiotics* (Eco, 1976; Klinkenberg, 1996). Semiotics offers not only a proper framework and language for describing this approach, but also the understanding and the directing of the design and development of hydroinformatics systems.

Working with and relying on the same semiotic methods, hydroinformatics must now assert its sociotechnical role through the analysis, design and development of new kinds of tools and environments that are much more suited to the *sociotechnical nature* of our current water-related problems. This means that the new tools should not only make it possible to increase the number of users of water-related electronically encapsulated knowledge, which is still predominantly scientific knowledge. The qualitative leap forward that is now proposed in these new tools is that they should provide means to mobilise this knowledge in such a way that it can be combined with other 'knowledge*s*', that also necessarily arise in given water–related context. There is an immense work ahead to bring a previously separated scientific knowledge under one unified framework with 'other' knowledge*s*, and especially to mobilise it so as to overcome another kind of separation, which is that occurring between people. The potential users of the new tools are for the most part not expected to be specialists in water related sciences, but are expected to include many other, more 'ordinary', citizens, on many social levels, who become involved in decision making about water related problems precisely by availing themselves of the means so provided.

Even though we will present these new tools in this study in terms of new kinds of decision support systems, it will be obvious that the processes of decision making are here understood only as *emergent* processes that result from the knowledge circulation that is enabled by the tools. In fact these new tools may be more properly described as platforms, or even electronic environments, that enable the encapsulation, expression and transmission of the different kinds of knowledge*s* *through* the interaction of the majority of those interested in the decision-making process. Such a broad discourse presents an overwhelming challenge for the designers of these new tools, because the requirements are now so much more manifold. This environment has to be, firstly, a device which can be used by many persons, usually simultaneously, so as to extract relevant knowledge both about the natural system and about the beliefs and positions of all the users who are involved in the decision making process. Secondly, however, at the same time this device has to be able to allow each user to express his or her *personal* interests and concerns to all the others. The ultimate goal is then that of providing a platform which will make it possible for individuals to engage themselves in negotiating and collaborating with other individuals in order to reach complying decisions. In fact, from a social point of view, such a device may be said to have only one inherent rule of its own as its only performative criterion, namely to induce habits of compliance with agreed solutions.

These requirements point in one definite direction. The design and the deployment of these systems cannot remain the *exclusive* task of their designers. The users of this environment will have to be continuously involved in the re-designing, transformation and the evolution of the tool itself. Even the boundary between the designers and users may become blurred. As Castells (1996, p33) explained:

> New information technologies are not simply tools to be applied, but processes to be developed. Users and doers may become the same...It follows a close relationship between the social processes of creating and manipulating symbols (the culture of society) and the capacity to produce and distribute goods and services (the productive forces). For the first time in history, the human mind is a direct productive force, not just a decisive element of the production system.

This should not only be understood as a tribute to and recognition of the key role of human knowledge *of any kind,* but also as a guide to the design and development of future systems for promoting knowledge circulation.

One aspect, which certainly must not be overlooked, is the recognition of the ongoing process of knowledge commodification. Electronic knowledge encapsulation and its dissemination have already contributed to the transformation of present hydroinformatics systems and tools into *products*. In fact the commodification of knowledge has become strongly associated with hydroinformatics. The field as a whole *already* marks the transition from knowledge as an attribute of 'knowers' to knowledge as a commodity that is encapsulated and marketed so as to be applied by 'consumers of knowledge'. It thus corresponds to the most predominant feature that marks the transition from a modern to a postmodern condition. The design and development of the new systems as they are presented here cannot be conceived outside of the realm of knowledge commodification. On the contrary, design and development have to be directed to using the opportunities of mass consumption of knowledge as much as possible. This aspect is closely related to the possibilities for introducing these systems into our societies.

At the end of this chapter we need to stress one characterisation of hydroinformatics that has been repeated several times since the inception of the field, and is particularly well suited to describing hydroinformatics as a sociotechnology. That is the notion that hydroinformatics always aims to occupy a *middle ground.* When hydroinformatics was initially introduced, this middle ground was understood by most people as a middle ground between the physical water sciences and ICT, as its name certainly suggests. We have tried here to introduce the further notion that hydroinformatics has another, third, pole which is just as important to it as the first two. That is the pole of the 'social', which from hydroinformatics point of view has much broader meaning than is implied in the usual use of the term. It is intended to convey the meaning that there must be an awareness of the need to *understand* as much as possible the social context within which hydroinformatics carries out its endeavours. It is only through such an awareness that hydroinformatics can in fact 'keep its feet' on the ground, which then becomes first and foremost the ground of the *ethical*. Thus the *middle ground* of hydroinformatics as presented in Figure 1.1 is at the centre of gravity of the triangle formed by these three poles. Because of its reliance on physical science, hydroinformatics has its *strength*, because of its employment of the ICT it can become *powerful*, and because of its social awareness its applications provide *value*.

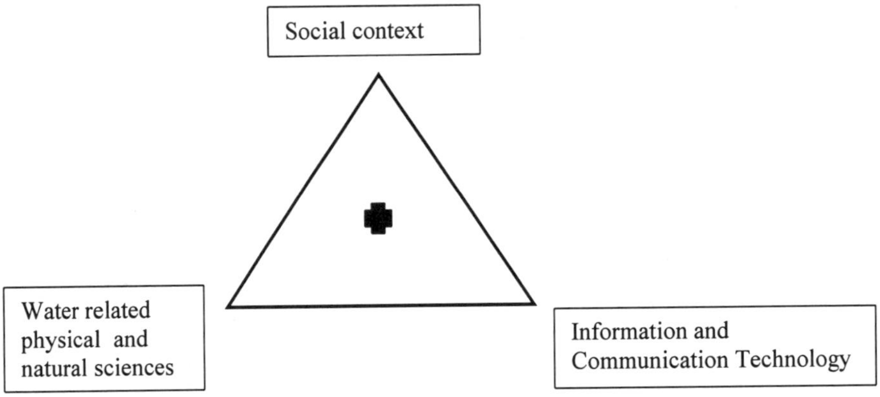

Figure 1.1 The *middle ground* of hydroinformatics

Thesis 8: *Hydroinformatics as sociotechnology recognises the hybrid nature of our current water and environment problems. In order to continue providing its benefits, hydroinformatics needs to take up the task of designing and developing new sociotechnical systems for knowledge circulation, while positively exploiting the ongoing processes of knowledge commodification. Hydroinformatics can only achieve this while occupying the middle ground between water-related physical sciences, ICT and the actual social context where hydroinformatics systems are being introduced.*

Chapter 3

Network Distributed Decision Support Systems (NDDSSs) in the Fields of Water and the Environment

Those very ones who believe that everything has been said and done,
will greet you as new and yet will close the door behind you.
And then they will say again that everything has been done and said
Eugène Delacroix

3.1 NDDSSs: Towards a new paradigm for decision making

From the previous chapter, we have seen that it is by virtue of the very nature of our current water and environmental problems that we are led towards the sociotechnical approach in dealing with the processes of decision making in these areas. Correspondingly, the analysis, design and development of the new kinds of computer and network-mediated tools and environments, including the new institutional settings, has to follow this same sociotechnical approach (Thesis 1 of Chapter 2). The diversity and the increasing complexity of the interests involved in our present water and environmental problems requires that we take a new perspective upon what are traditionally known as decision support tools and systems. Instead of thinking about tools and systems that are developed to be applied by a relatively few 'decision makers', there is a need to develop decision support *environments*, where all interested parties and stakeholders, including individuals, can participate. This means that the traditional understanding of 'decision support' will have to be taken beyond the development of various optimisation modules developed by expert decision analysts who translate the decision making problem into some sort of optimisation problem and present the solution of this problem as a recommendation to the decision makers. Even though these optimisation modules continue to be built, and even with a constantly increasing complexity, they are rarely capable of capturing the actual social environment within which the decisions are to be implemented. Therefore, the primary concern in the development of the new support environments becomes the inclusion of all interested parties in the decision making process as opposed to building complex optimisation models. In fact there is a realisation that it is far more important to provide *satisfying* solutions than it is to generate *optimal* solutions. The optimal solutions resulting from no matter how complex and all-inclusive optimisation procedures remain optimal only *from a particular point of view*. Without direct participation of all interested parties, this point of view is in practice very often concerned with the interests of the traditional centres of decision making, usually composed of various corporate bodies representing industry, the market and the government.

In fact, this shift in the understanding of the nature of the decision making processes, and the corresponding development of devices for realising these processes, are together so important, that we may speak of *a new decision making paradigm*. The common distinction between the current, traditional decision making paradigm and the paradigm that arises on the basis of this new understanding, can be expressed in terms of 'top-down' and 'bottom-up' approaches. Even if this description does not cover all the requirements of the new approach, it is well suited to an initial introduction of this paradigm shift. Using this description, the two paradigms can be presented as shown in Figures 3.1 and 3.2.

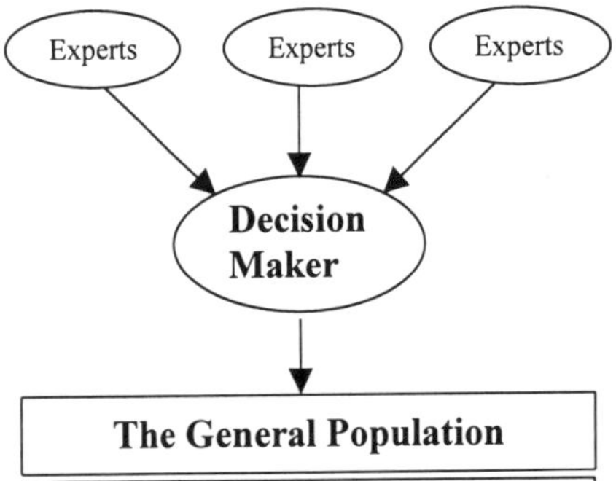

Figure 3.1. The current decision making paradigm (adopted from Abbott, 2001)

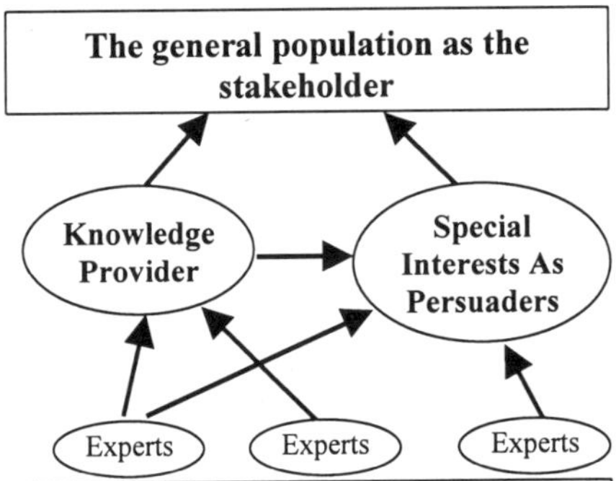

Figure 3.2. The new decision making paradigm (adopted from Abbott, 2001)

In the established order of the current decision making paradigm, the actual decisions are always made within the social domain, and the responsibility for these decisions is taken by people commonly known as decision makers. However, because these decisions are nowadays so much dependent on technology (measurements, modelling results, etc.) the need always arises for various expert support that will summarise the multiple aspects of the decision making problem and present the decision makers with a clearer picture of the available options. The general population, as well as many stakeholders who are excluded from this process for whatever reasons, usually receive the decisions made as already made 'solutions'. The request for their compliance with these decisions is usually supported by the argument that these solutions correspond to the result of 'expert analyses'.

It has already been realised in many places that this approach is no longer tenable and that the inclusion of all the genuine stakeholders, included interested persons in the general public, is indispensable in the decision making process. This realisation is spreading especially in water- and environment–related problems, where human interventions may have far reaching consequences for so many different kinds of stakeholders. Even while this is being recognised, however the tendency remains for the most part to try to accommodate all these different interests *within the existing paradigm*. The expert advice is thus not only restricted to the various technical aspects of the decision problem at hand, but the analysis of the actual social domain where the decisions are to be implemented becomes also a task to be carried out by 'the experts'.

The first attempts to overcome problems in this area while still following the 'top-down' approach were characterised by building tools on the basis of so-called multi-criteria analyses. Although the emphasis was here more on the inclusion of the various interests than on the optimisation process itself, in practice the implementation of the outcoming results was still problematic. Behind these approaches remains the philosophy of the corporate-centred decision maker with one (inherently false) assumption, namely that if there are various interests that should be included in the decision making process it is possible to foresee and describe them in advance. The experts, working on behalf of the 'decision makers' shared this conviction and even extended it by the inclusion of the foreseen interests and intentions of the various stakeholders in the design of the corresponding decision support tools and systems. Although various methodologies for identifying and foreseeing all manner of different interests have been proposed, the process remains fixed within the same framework, since all these interests are in the end being *translated* by the experts into a formalised decision problem. Even if it appears that this is directed away from that technocratic thinking which underlies the current decision making paradigm, the governing function of the 'efficiency' criterion remains inherent in all these processes and can rarely be questioned (corresponding to Thesis 2 in Chapter 2).

Observations of the repeated failure of this approach have gradually increased an awareness that what is needed is to involve "all the stakeholders" *directly* in the decision making process. 'Public participation' has been identified as crucial in any water-related decision-making process (World Bank 1993). At first this was only a formal statement of desideratum, not backed by any real commitment, but even when later the commitment came to be present it appeared to be extremely difficult to translate it into practice. Usually it becomes 'too expensive' or otherwise 'technically unfeasible'.

The proposal that is advanced here for overcoming this situation is to view these problems as one subset of a more general set of problems of *knowledge circulation*, and thus as problems which can only be resolved through the introduction of new kinds of sociotechnical tools and arrangements. In fact, the newly proposed paradigm, as represented in Figure 3.2, has regarded as an objective that may be attained through the introduction of the newly propose virtual environments. As seen from the figure, the roles of all the participants in the decision making process are still present, but now within an inverted framework. While the role of the general population now is at the top, indicating their increased, direct participation, the role of the traditional decision makers is transformed into a generic role of 'persuaders'. These decision makers, who are usually representing corporate interests, are now embedded into a much wider environment. In fact the systems proposed are aimed simultaneously at increasing the awareness of the corporate interests also about this wide environment. In this situation, in order to satisfy their goals the corporate entities become necessarily involved through their own active participation in the processes of knowledge circulation. Their engagement becomes primarily one of negotiation and collaboration with - and persuasion of - all the others involved in the process. The new component in the paradigm is the knowledge provider, which has to be understood here as a generic name for the totality of the sociotechnical infrastructure which in part provides - and overall manages - the knowledge 'content' that is brought into circulation. This component is actually the central topic of this study. The role of the experts is maintained in the new paradigm, but instead of their efforts being directed to supporting only 'the institutionalised decision makers', their expertise becomes much more important in supporting the activities of the knowledge provider, and thus ultimately of a much wider range of participants in the decision making process.

Although in the fields of water and the natural environment an awareness is spreading that the shift towards the new paradigm is necessary, the task of engineering this shift will remain difficult, and probably even unattainable, without the introduction of new kinds of sociotechnical arrangements developed specifically for this purpose. The name that we give to the new sociotechnical constructs that will serve to support the new decision paradigm is Network Distributed Decision Support Systems (NDDSSs). The word 'network' refers here primarily to the fact that these new sociotechnical systems have to be developed and deployed over electronic networks such as the Internet. Thus, the electronic networks are only technologies that enable the realisation of these essentially *socio*technical systems. What the systems themselves enable is the creation of networks of people and institutions, engaged in processes of knowledge circulation.

As we have already stated (Theses 3, in the preceding Chapter) one of the main goals of these systems is the synergetic mobilisation of as much scientific and narrative knowledge as possible for the purposes of decision making. This inevitably means that these new systems have to cope with such a broad decision making discourse that the traditional approaches oriented towards the formalisation of decision making problems become too restrictive. These new systems simply cannot be "tools to be applied, but processes to be developed" (Castells, 1996), and this imperative has to remain as one of the principle guidelines in their design and development. Nevertheless, it is still necessary to define, even if in very broad terms, certain 'design requirements' for these systems. The development of particular instances of these systems may frequently follow different paths, but since this study is dealing with the basic introduction of these systems, the broad requirements still need to be clearly stated.

3.2 Design requirements for an NDDSS

Although we speak of a continuous knowledge circulation within these systems, we can envisage that this process proceeds along a complex criss-crossing network of paths or channels of knowledge transmission. It is therefore appropriate to link the design requirements for NDDSSs with the principle directions of knowledge transmission within this network of paths or channels:

1. NDDSSs have to enable the access to relevant parts of the body of scientific knowledge, primarily those pertaining to the effects of interventions in the physical domain, within a given decision making context. This access should be provided in such a way that various alternative actions relevant to the problem at hand can be tested by any user-participant independently.

This access has to be provided to all participants in the decision making process, regardless of their position in any kind of social hierarchy (Thesis 3 of Chapter 2). Because of the perceived diversity in the background knowledge, the values and the interests of the participants in the process, the same body of scientific knowledge has to be made accessible in various *forms*, as well as on various temporal and spatial scales.

2. NDDSSs have to enable each participant in the decision making process to express his or her interest and intentions, regardless of whether the participant represents corporate interests, individual interests, or the interests of groups of individuals.

Many interests that are already established in a given community as 'common interests' for large parts of that community, as well as the identified corporate interests, should already be accommodated in the initial design of the systems. The systems, on the other hand, have to be designed as *open* systems in the sense that they must be able to accommodate (or evolve to accommodate) other interests, and then primarily individual interests and concerns that cannot be envisaged beforehand. As a consequence of this:

3. NDDSSs should allow all the participants to combine the relevant scientific facts about the proposed changes in their physical world with their own beliefs and intentions in order to create their own judgement of the proposed and the courses of action that they consider more desirable.

This is one of the most difficult tasks to be realised, especially because, despite some thousands of years of study, the judgement-forming processes that come into play are still far from being sufficiently understood. All relevant, expert knowledge that can assist users in this process has to be provided, while at the same time resisting the inherent tendencies of 'locking' the judgement-forming process into formalised procedures designed around that same expert knowledge. The judgement-forming process must in fact be allowed to proceed along lines that are *not* envisaged initially: precisely because the participants may be aware of specific local conditions, they may hold different beliefs and attitudes, or even have completely unforeseen reasons for a particular judgement.

For corporate interests that are forming their judgements on the basis of their own kinds of decision support tools or economic models, an interface has to be provided to these sources of scientific knowledge, which thereby become incorporated into the system.

4. NDDSSs must provide transparent presentations of all the different interests involved, even if the number of these interests is very large.

In particular, those issues that are already established as 'common concerns', or are gradually evolving into a more widespread acceptance as 'common' have to be presented to all the users. In this way the users can form their own judgement not only against the background of their own position and the relevant scientific knowledge about the changes in the physical world, but also against the background of those issues of common concern that have already been raised. The evaluation of the proposed courses of action must be allowed to proceed for each participant's choice of issues of concern, as driven by their individual situations and intentions.

This issue, which is essentially one of transparency, must also be understood in terms of the 'publicity' of the actions that any participating party undertakes during the whole decision making process. It is one of the key qualities of these systems that has to be introduced, both for promoting the changing of attitudes among the participants, and inseparably from this, for promoting their processes of secondary reflection (Thesis 4 in Chapter 2). Then:

5. NDDSSs need to provide appropriate procedures for aggregating all the different judgements and evaluations, carried out by the participants in the decision making process.

This is a crucial step for initiating the processes of negotiation and collaboration that should follow. This aggregation should provide a picture of the 'social landscape' in the particular decision making context. When realised, it will allow all the participating parties to view and compare their own position against the various aggregated reactions of the whole community. Thus:

6. NDDSSs need to serve as platforms upon which succeeding processes of collaboration and negotiation can be supported, with the ultimate goal of reaching complying decisions.

This is again a complex task that cannot be easily realised. Similarly to the judgement forming processes, the collaboration and negotiation processes are not amenable to formalised treatment. The main task of the system in this respect is that of supporting the 'social learning' process, for which the quality of transparency is of greatest importance. By way of providing a direct support to the negotiation and collaboration processes, the system has to facilitate coalition and alliance building processes, and therefore provide support to participants in the process in exploring and suggesting options for realising mutual advantages.

These are indeed vary broad design requirements, and they are certainly not exhaustive. However they should serve as general guidelines for the initial, conceptual design of NDDSSs. It may appear that NDDSSs are simply tools that should support the steps commonly required in any approach that is genuinely dedicated to the inclusion of all stakeholders: *identification, evaluation* and *communication.* Indeed there are elements of all these steps in an NDDSS. However, with the number of participants – potential users of such environments - envisaged to be many thousands, and the requirements stated above, it should become clear that NDDSSs cannot be seen simply as tools that formalise *procedures* to be followed. Instead, NDDSSs are knowledge circulation environments that should enable *emergent decision making processes* (Thesis 8 of Chapter 2).

From the above stated design requirements and from the general objective of NDDSSs we can recognise a connection between the analysis, design and development of NDDSSs and the field of Environmental Impact Assessment (EIA). Most of the requirements actually correspond closely to the required steps in the established procedures for carrying out EIA studies (see for example, Wathern, 1992; Bregman and Mackenthun, 1992; Pawelas and Mostert, 1994). In order to point out more clearly the areas where the field of EIA is compatible with the NDDSSs design and development, and to identify the ways in which they can reinforce each other, we need to take a brief look at the changing nature of the EIA process itself.

In their present state of development, EIA systems are being established as *procedures* which are increasingly accepted in many countries of the world through the adoption of corresponding institutional, legal and judicial frameworks (see, for example, Sadler, 1996; European Environmental Bureau, 1998; or, The Regional Environmental Center for Central and Eastern Europe, 1998). These procedures are now at the core of EIA systems and processes at the level of project-related EIA. Similar procedures are gradually becoming established in the emerging field of Strategic Environmental Assessment (SEA), which refers to application of EA to policies, plans and programmes, whereby environmental considerations be brought to the highest decision making levels. In these procedures the issue of public involvement and participation is repeatedly stressed as a key requirement for an effective EIA process. In particular, the need is recognised for moving beyond informing and consulting and towards meaningful participation. As expressed by Pawelas and Mostert (1994, p. 72):

> Public participation in EIA should be seen as the most appropriate means to achieve democratic values, particularly the representativeness and responsiveness of political and administrative decision making, rather than as a means to fulfil a regulatory requirement in the assessment.

Many methods have been proposed within the EIA procedure for achieving meaningful participation, such as surveys of stakeholder expectations, public consultations, public communications and awareness-raising campaigns. On many occasions, similar problems to those of NDDSSs design have already been experienced. For example, the issue of promoting *intersubjectivity* has already been recognised at the level of the multidisciplinary expert panels involved in the early stages of the EIA process, such as those directed to screening and scoping. At the level of stakeholder and public involvement, the same problem occurs even more in the later, evaluative stage of the process.

In fact, most recent recommendations call for stakeholder and public participation *throughout* the EIA process, and so in all of its stages. It is already apparent, however, that the established EIA procedures become significantly less appropriate for dealing with the resulting manifold increase in the complexity of the process. Even the established legal and institutional frameworks, coined in international conventions, fail to meet the demands of such direct participatory process. Palerm (1999), for example, identified four fundamental aspects related to public participation that need to be incorporated in future EIAs:

- the need to ensure the participation of cognitively and lingually non-competent actors;
- the need to have a two-way communication process;
- the need to ensure that normative and subjective claims are adequately recognised;
- the need to establish conflict management procedures;

With the envisaged large number of individuals constituting the 'public', with their diverse beliefs, attitudes and interests, it is obvious that the existing tools and methodologies cannot achieve the above stated goals. If we examine the design requirements for an NDDSS, we can recognise that it is exactly in these areas that these new sociotechnical constructs can offer their major contribution. From one side, the initial NDDSS design can benefit from the existing structured procedure already established in the field of EIA, while, from the other side, NDDSS themselves can become means for achieving truly participatory EIA process.

At the same time, the nature of EIA is changing in the direction of much more integrated processes. A so-called second generation, multi-level, integrated approach to EA is emerging which builds on the formal procedures of EIA, together with those of social impact assessment (SIA) or risk-assessment (Sadler, 1996). This integrated approach necessitates the establishment of various linkages to national planning and decision making processes that depend in turn on the existing institutional framework and management, as well as upon existing customs and traditions. In fact it is probably more appropriate to talk of incorporating the EIA procedures, and especially SEA procedures, into the much broader area of *Policy Analysis.* It is in such an integrated framework that the linkages between EIA and decision making become crucial. Because NDDSSs are envisaged not only to act as systems that enable the participatory impact assessment, but as means to influence decision making processes directly, the overlaps between NDDSSs and the field of policy analysis need to be also identified.

As a first step in this identification it needs to be recognised that field of policy analysis is already undergoing changes which are influenced by our postmodern condition of society. In particular, research in policy analysis has already demonstrated the problems involved in extending the traditional use and application of scientific knowledge to policy making (see, for example, Majone and Quade, 1980; Weingart, 1999; Bruijn and Heuvelhof, 1999). This research clearly points to the problems related to an 'inflationary use of science', which only reinforces the role of experts in policy analysis and makes it more difficult than ever for the broader public to have an influence. It also identifies certain paradoxes associated with the 'scientification of politics which results in politicisation of science', and the general 'delegitimation of scientific knowledge'. It can be argued that all these problems are fundamentally related to the separation between scientific and narrative knowledge, as discussed in Chapter 2. As the result of the realisation of these problems, the field of policy analysis is being transformed, at least at the level of theoretical analyses, in a typically postmodern manner. These transformations, which have accelerated over the last two decades of the twentieth century, have resulted in a number of different streams of theoretical policy analysis, such as 'relativistic policy analysis', 'dialogical policy analysis', 'critical-theoretical policy analysis', or 'forensic policy analysis' (see, for an overview Hoppe, 1999). Increasingly, policy analysis sees its most important role in mediation and advice, avoiding the use of 'mandated science' and facilitating 'negotiated knowledge' (Bruin and Heuvelhof, 1999). This general transformation in the theory of policy analysis is usually described as the 'argumentative turn', or 'post-positivist turn' in policy analysis (White, 1994; Durning, 1999). Admittedly there is still a large gap between these new theoretical concepts and the practice of an ongoing policy analysis which still largely follows the traditional approach.

Of significant importance to this study is the recent identification of policy analysis as a 'participatory policy-oriented process' (Thissen and Twaalfhoven, 2001). Although this is only one of the views expressed in current policy analytic studies, it is one that closely corresponds to the study of NDDSSs. In particular, its theoretical basis (which identifies a number of *actors* involved in a policy *network*, with interdependencies among the actors, the variety of actors' goals, and the resulting relation patterns among them) can be beneficially used for the conceptual design of a given NDDSS. This approach again stresses the *intermediary* role of participatory policy analysis (Thissen and Twaalfhoven, 2001):

> From the perspective of policy network theory, policy analytic activities have a role as an intermedator, supporting communicative debates among multiple parties at interest, developing shared strategies, creating win-win situations, and breaking through cognitive fixations...A logical consequence of the network perspective is that policy analytic activities should not aim at rational advice given to a single authoritative public decision maker, but should start from, and end with, the interactions between relevant stakeholders.

On the other hand, similarly to the situation that we have described in relation to EIA, fully developed NDDSSs can serve as the most appropriate *means* for achieving this role of policy analysis. Although some policy analysts tend to doubt the potential of technology, and especially of advanced Information and Communication Technology (see, again, Hoppe, 1999), which is understandable if the processes involved are approached from a purely technical standpoint, the *sociotechnical* approach, which is the basis for the design and development of NDDSSs, seems to be the only way forward towards the goals of participatory policy analysis.

The fundamental change that is required from the field of policy analysis in this sense is perhaps best summarised by Wildavsky (1993, p. 277), one of the leading theorists in the field:

> The rationale for analysis – avoiding experience by imagining what might happen if things worked this or that way – may be its ruin. The temptation of the analyst is to treat citizens as objects. By depriving people of autonomy of thought (their consciousness is false, their experience invalid), it is possible to deny them citizenship in action. The moral role of the analyst, therefore, demands that cogitation enhances the values of interaction and not become a substitute for it.
>
> Why is citizenship important for policy analysis? Because ... without participation by citizens:

1) capacity for enhancing autonomy by learning will decline;
2) damage will be done to human dignity by denigrating reciprocity;
3) therefore, many difficulties won't improve as much and as quickly as they should, and other will get much worse, faster and necessarily.

> The task of analysts ...[is to develop] a system of criticism which permits policies to be tested in such a manner as to enable a self-correcting capacity. And...this testing cannot and should not be done only by policy analysts, but also (and primarily) by policy participants. The less necessary analysts make themselves, by enhancing the self-correcting characteristics of the relationships within a policy, the more they have succeeded. The moral trap of the professional, as Max Weber knew, is "living off" instead "living for" the clients. Doing oneself out of a job is the best way to do analysis.

The 'doing oneself out of a job', is represented exactly by the changed role of the experts as depicted in Figure 3.2. Instead of providing prescriptions, they become advisors, persuaders and mediators. Both within the design and development of NDDSSs, and in their future use, their new role remains indispensable.

In presenting the broad design requirements of NDDSSs we have tried to link the tasks of design and development of these systems with the fields of EIA and policy analysis. While the overlaps with these broader fields are apparent, and the proposed approach has a correspondingly broader potential, in the following presentation of the concept of NDDSSs and of its functional components, we shall focus primarily on water-related problems as the most immediate problems for hydroinformatics.

3.3 The general concept of NDDSSs

Let us see now how we can conceive of a system like an NDDSS. Since we are here concerned with water-related problems, an NDDSS is conceived as dealing with decision problems related to a particular spatially bounded object such as a river basin, a coastal region or a groundwater system of aquifers, or even some or all of such features combined. The environment provided by the NDDSS can be conceived at the simplest and at the same time most general level as presented on Figure 3.3. At the core of this environment is a so-called 'knowledge centre' which is the primary repository of all what we consider as scientific knowledge. This is primarily electronically encapsulated knowledge, and first and foremost this means various models (hydraulic, hydrological, ecological etc.). The knowledge centre is also the repository of all relevant data that are needed either for generating or supporting these models. In many cases we may conceive that the knowledge centre will have to maintain continuous links with all kinds of measurement facilities (both from data gathering equipment on the ground and from remote sensing data such as satellite imagery, radar data for rainfall prediction etc.). Although the gathering and distribution of this kind of knowledge is the primary role of the knowledge centre, other kinds of knowledge that are not directly related to water sciences should also be located here. For example we can envisage that on top of this part of scientific knowledge, any complete centre will eventually have to encapsulate legal knowledge bearing upon the problems in question. The users of the system may now be visualised as being placed on a ring around this knowledge centre. These are both corporate bodies with their own perceived interests, as well as individuals or groups of individuals grouped around some common concern.

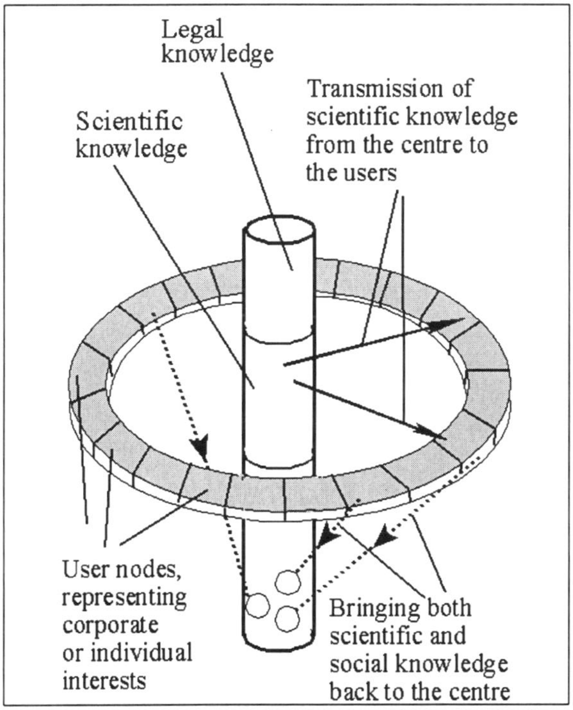

Figure 3.3 The general concept of NDDSS (adopted from Jonoski and Abbott, 1998)

It is very important to stress that the knowledge centre should not to be seen as a physical centre of the system, or, even worse, to assume that this is a centre only in technical terms and that it can be reduced to one machine acting as a server. The word 'centre'[9] is used here only conceptually, and the models and data provided by this centre might well reside at different physical locations, and, quite commonly, in different organisations. Achieving proper connections and interoperability of all these repositories of scientific knowledge will therefore require not only resolving a number of technical problems, but also introducing adequate institutional arrangements. The understanding of the knowledge centre as an 'unmanned' part of the system, which can easily be put into operation after all technical problems have been resolved, is almost certainly unrealistic.

Once we visualise the environment in this way, we may say that it has to provide a transmission of knowledge in two directions. The first of these is from the centre outwards, whereby scientific knowledge is brought to all the different users. This first direction of knowledge transmission is immediately associated with a number of problems, some of which still have to be resolved. Firstly, the form in which this knowledge will be brought to these different users may vary considerably. For example, some of them will request spatial information presented in different kinds of maps, most probably relying on GIS facilities. Others will request information presented as simple numbers, or graphs or tables. More

[9] This conceptualisation itself, with the repository of scientific knowledge at the 'centre' and all the users somehow at the 'periphery', is certainly 'modern-centric'. An apology for this bias is perhaps appropriate, but the reader is asked not to understand this conceptualisation in the sense of assigning the *dominant* position to the knowledge coming from the centre. The 'centrality' of this knowledge should be understood only in terms of its relatively higher amenability to formalised treatment, as compared to the diverse knowledge*s* available at the user nodes.

traditional textual descriptions will also have to be provided for some kinds of knowledge that is delivered through the knowledge centre. For the users that are far from our water related sciences, the most appropriate form may be some sort of cartoon presentation of the expected consequences of proposed actions. Another problem that comes immediately into the picture is that different users will require results on different temporal or spatial scales. Thus, there is already a complex of problems, associated with this first direction of knowledge transmission alone, from the centre outwards to the users, which will require considerable design and development efforts.

However, the second direction of knowledge transmission opens up a completely different set of problems. At the user nodes the users/participants will firstly have to combine the scientific knowledge which they have obtained from the centre together with their own 'social knowledge' in order to form judgements about most of the proposed courses of action. The diversity of the social knowledge available at the user nodes will then present even more formidable challenges to the design of the system components that will support this judgement-forming process.

Secondly, this combining of scientific and narrative social knowledge has to be extended with knowledge about the attitudes, positions and interests of other participants in the process. This knowledge then becomes a precondition for the participation of all users/participants in the processes of negotiation and collaboration that may in most cases ultimately lead to complying decisions. In order to be able to become engaged in collaboration and negotiation with the other participants, they will have to know who these other participants are, and to know what they believe and know, so that they can reason not only with their own knowledge but also with the knowledge of the others participating in the process (thesis 4 in Chapter 2). Without significant assistance provided by the system for this purpose, these tasks will be next to impossible for any human participant. Even with only a small number of users - with possibly conflicting interests - the complexity of relations that may develop is difficult to handle. With the large number of users envisaged as potential users of an NDDSS, this task seems impossible unless components are designed and oriented specifically towards providing this kind of support.

With this very general description of the concept of an NDDSS, we have gradually identified the principle processes that a system of this kind should support. These principle processes in turn determine the specification of the main functional components of an NDDSS. We shall now introduce these functional components because they are seen as necessary for any NDDSS, even though different systems will most likely employ different approaches towards the realisation of these components.

3.4 Functional components of NDDSSs

There are three main functional components in an NDDSS: the *fact engine* component, the *judgement engine* component and the *platform for collaboration and negotiation facilitation.* These components are located in the two distinct conceptual parts of the system: the knowledge centre, and the users' periphery (Figure 3.4).

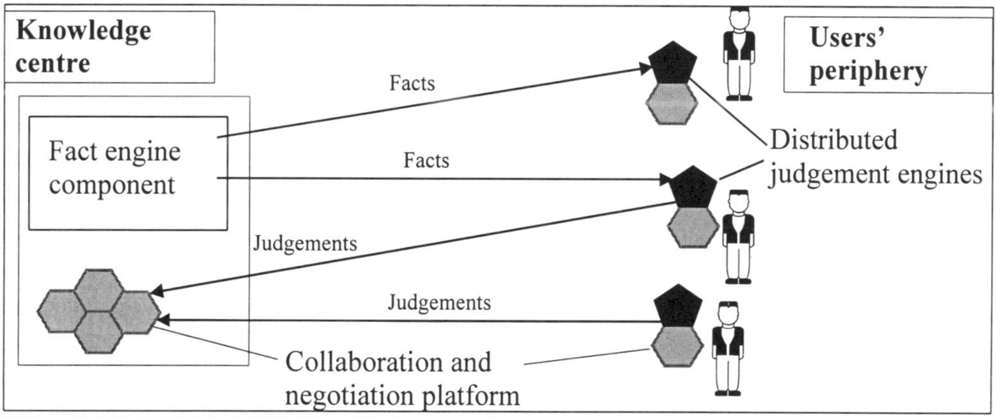

Figure 3.4. Three functional components of an NDDSS

The fact engine component can be seen as being almost entirely located in the knowledge centre, while the judgement engine component is in this context distributed over the user nodes at the users' periphery. The platform for collaboration and negotiation, on the other hand, has components that are overlapping both with the fact engines and with the judgement engines and can be regarded as being distributed between the users' periphery and in the knowledge centre.

3.4.1 Fact engines

The *fact engine component* is in principle located in the knowledge centre even though many knowledge providers and nearly all data sources may be physically situated elsewhere. This component primarily provides the storage and access to a collection of models and data related to various aspects of the physical system in question. Most of these models will initially be traditional numerical models, such as various hydraulic, hydrological, ecological or water quality models. Since the main role of these models is to provide the *facts* about the physical system in question, we classify them under the category of *fact engines*. The access to these models is now necessarily going to be through the Internet, or some similar kind of electronic network. Because of this, the first thing that comes to mind is that the traditional user interfaces, designed for their use as stand-alone applications will have to be de-coupled from their computational engines and embedded in other media, such as, for example, a Web-browser. (This is a direct consequence of Thesis 5 in Chapter 2)

However, the much broader potential user base of these new systems indicates that these models may need to be stripped from their traditional user interfaces and replaced by new ones, which will be more adaptable to the diversity of the users. In fact, alternative kinds of user interfaces must be provided that will be able to represent the *same* facts in *different forms*. Because this aspect is actually closely related to the judgement-forming process that occurs at the user nodes, and is to be handled by the judgement engines, the user interfaces within an NDDSS are *not* primary considerations in the design of the fact engine component. The existing user interfaces will certainly be indispensable for setting up, or 'instantiating' the models as fact engines – a task to be carried out by the modellers who are working at the knowledge centre. However, the interfaces to these instantiated models that will be available at the user nodes will need to be reconstructed in such a way that their primary goal is to enable the judgement-forming process, and *not* the modelling process per se. This firstly implies the representation of the modelling results in different forms, that will support the diverse backgrounds and knowledge of the users, but it also places limitations on the number of model parameters or inputs that can be manipulated.

The diversity of the user needs, however, points towards another important aspect directly related to the design and development of the fact engine component. Traditionally, and in full accordance with the established modern approach or tradition, *separate* models have been developed for separate kinds of water-related problems. Hydraulic, hydrologic, water distribution, groundwater or water quality models all deal with separate kinds of problems. Within each class of these models there are numerous variations in terms of spatial and temporal scales, as well as underlying assumptions, and indeed there are classes of models that are particularly suitable for particular classes of problems. Only more recently has much effort been put into providing *integrated* modelling systems which try to combine as many models as possible into one conceptual and computational framework. Beyond this again, seamless interfaces between modelling systems and Geographical Information Systems (GISs) have already been extensively developed as means of integrating water-related results to corresponding geographical circumstances.

Within the framework of an NDDSS, *integrated modelling* gains an importance that proceeds beyond the need for providing more relevant, reliable and accurate results. The large number and the diversity of the needs of the users of the system require the introduction of increasingly integrated modelling approaches. An intervention in one part of the physical system in question can have such far reaching consequences to other parts (and this is meant both in spatio-temporal terms and in terms of different domains), that many modellers are already aware of this aspect. While in the traditional approach it was still often possible to ignore some of these consequences, within NDDSSs, that are oriented towards satisfying as many knowledge needs as possible, integrated modelling becomes indispensable.

There are many developments towards integrated modelling, and some existing modelling systems are already employing quite integrative approaches, but it needs to be realised that the design and development of NDDSSs will require a level of integration that is on a completely different level again. Water movement in all different phases of the hydrological cycle will increasingly have to be combined with water quality models, models of urban networks for drinking and waste-water, and models of water and waste-water treatment processes. All these water–related models will then have to be combined with ecological and other environmental models. Even if a general framework of this integration is inconceivable outside of a given decision making context, the development of the fact engine components of NDDSSs will be greatly facilitated by integrated modelling systems.

Next to a new level of integrated modelling that provides linkages of knowledge across separated water- and environment-related domains, and on different spatial and temporal scales, it is important that the knowledge about the effects of the proposed courses of action is provided at a suitable, and sufficiently high, resolution. It can be expected that the real value of a given NDDSS to an individual user/participant will depend upon the possibility thereby provided to follow effects on certain features that correspond to his or her individual interests, such as the flooding of an individual house or street, or the lowering of the groundwater table at an individual field. While for many other corporate participants this level of resolution may not be that relevant, bringing NDDSS to the level of individual participants will be crucially dependent on it. (For an example of one development in this direction, related to flooding problems, see Martin and O'Kane, 2000)

It has to be realised, as well, that the traditional numerical fact engines are not the only ones that will be used in systems of this kind. In fact some of the serious drawbacks of these traditional models (as primarily related to their computational efficiency) will require their replacement by different kinds of models, in particular those of the so-called sub-symbolic class, such as are realised through the construction of neural networks or through genetic programming (e.g. Dibike et al.; 1999; Minns, 2000; Babovic and Keijzer, 2000), or other models more generally described as 'data-driven' (e.g. Velickov and Solomatine, 2000). The advantages that these models offer in terms of fast responses to given inputs may be very important for systems that have to provide results to very large number of users over the Internet. We can then speak of neural-network-type fact engines or genetic-programming-type fact engines and indeed of other kinds besides. In any case these will still belong to the general category of fact engines, since their *function* as fact providers remains the same. As will be discussed later, these data-driven models and the whole field of *data mining for knowledge discovery* in general can be exploited in the other components of an NDDSSs as well, and potentially even more beneficially.

The water and environment-related models, however, are not the only fact engines in the fact engine component. Many participants in the process may use different fact-providers such as various databases (often containing aerial photographic and satellite data), knowledge bases, expert systems, or even simple texts, some of them conveying knowledge that is not directly related to the specific physical system in question. This means that the development of fact engines will *generally* depend on achieving *software interoperability*. This problem already arises at the level of integrated modelling, of course, but, in the light of the existence of other fact-providing software, the extent of this problem becomes even clearer. A fact engine component of a particular system may contain dozens or even hundreds of different individual fact engines that would have to work with large dependencies on each other. Achieving their interoperability is then a formidably difficult task. Again, general recipes for achieving software interoperability do not exist and various technologies may be employed for different tasks and domains. One framework based on a so-called *agent orientation* has been recently proposed, where different software components are being conceptualised and implemented as *software agents* able to communicate with high-level communication language. This approach will be addressed in chapter 5.

3.4.2 Judgement engines

The *judgement engine component* will usually and primarily be distributed over the users' periphery, where we may envisage that all the users who will participate in the impact assessment process and the decision-making process will be situated. The judgement engines have the task of assisting the users in making their own judgements about a particular intervention in the system, based on the interaction between their beliefs (or interests) and the received facts from the knowledge centre. The processes of building judgements based on the user's belief system through the intervention of facts can obviously be exceedingly various and complex. The lack of a sufficient understanding of these processes makes the design and development of the judgement engine component one of the most difficult tasks in the overall development of NDDSS, and one that still requires a great deal of fundamental research.

Because of the diversity of the users/participants involved in the process, the first prerequisites in the formation of judgements are related to the ways in which the facts coming form the knowledge centre are delivered. As we have already mentioned, the user interfaces of the fact engines may have to be significantly adapted for this purpose. The main orientation in the new design must be towards *customisation* and even *personalisation* of the content delivered from the fact engines. This customisation is related both to the *kind* of facts that are relevant to a particular user for making his or her judgement, and to the *form* of delivery of these facts. The design of these interfaces should especially be oriented directly towards assisting users that are not directly familiar with the water- and environment-related scientific knowledge in their judgement-forming process.

The delivery of the relevant facts is however only one step in the judgement forming process. The main role of the judgement engines should be that of the *combining* of the facts with the users' beliefs and interests in order the users to form their own judgements. It should be realised that the judgement engines of an NDDSS are envisaged as primarily supporting and facilitating the formation of judgements of *value*. The diversity in values and in value gradients of the users/participants requires that the judgement-forming process becomes understood within a framework where this diversity is respected and the tendencies towards restriction of certain values, or towards reduction of all values to one (usually monetary) value, are discouraged. While there have been many attempts directed to formalising these processes within the *rationalistic framework*, such a purely technocratic approach is almost surely inadequate for the envisaged purposes of NDDSSs. Although various corporate interests that will be present in the decision-making process will most likely follow formalised judgement-forming procedures, the inclusion of the individuals, and their own interests and intentions clearly delimits the applicability of this approach.

Because of the requirement to approach these processes within a much broader framework, a deeper theoretical insight into judgement forming processes *in general* is necessary. A fully developed theory that can offer clear guidelines for the design and development of judgement engines still does not exist largely because the vast literature on this subject has not been researched within the present context. Our approach towards the judgement-forming process must however necessarily start out from the *phenomenological* viewpoint, which sees all judgements being mediated by immediate experiences, or *phenomena*. From this viewpoint, such a full-fledged theory within the present context, even if developed, will not have any *direct prescriptive* power for the design of judgement engines. However, the *descriptive* power that can be mobilised by employing the *phenomenological method* may greatly improve our understanding of the judgement-forming processes, and at the very least point towards the right directions for future development.

Let us see then how the judgement forming process within an NDDSS can be approached from this viewpoint. We can imagine that different users/participants in this process will have different beliefs, attitudes, desires and intentions. While they will be forming their judgements on the basis of the *same facts,* as provided by the fact engines, their judgements about the proposed courses of action will depend on these cognitive features that are known under the common name of *mental states.* We cannot discuss the nature and origin of the mental states, which is a topic in such diverse fields such as philosophy of mind, cognitive science, artificial intelligence, and recently even software development. The introduction of mental states is however necessary for an *understanding* of the phenomenological approach towards the study of the processes of making judgements.

The primary concept that has been introduced for dealing with mental states is the one of *intentionality*[10]. This notion of intentionality is used to describe mental states in one particular sense that *all intentional* mental states are *about* certain objects or states of affairs. In this sense mental states are *directed* towards objects. This notion of *aboutness* of the mental states is very important, because it links the *intentional agent* (the human mind) with different kinds of objects. As described by Searle (1983, p.1):

> Intentionality is that property of many mental states and events by which they are directed at or about or of objects and states of affairs of the world. If, for example , I have belief, it must be a belief that such and such is the case.; if I have fear, it must be a fear of something or that something will occur; if I have desire, it must be a desire to do something or that something should happen or be the case; if I have an intention, it must be an intention to do something.

Apart from the 'aboutness', the notion of intentionality refers to mental states that cannot be considered in isolation. Mental states of any person have to be seen within the so called *intentionality network* (Devlin, 1995). Many mental states (different beliefs, for example) and many different *kinds* of mental states (beliefs, intentions, desires, for example) are maintained at the same time.

Within this framework of intentionality, the formation of judgements is a product of *intentional acts*, guided by a number of different mental states that together constitute a particular intentionality network. Intentionality is thus at the core of the *phenomenological method*, which aims to describe, among many other phenomena, the formation of *any* judgement. In more basic terms, this whole approach views the process of *abstraction,* which is the primary step in any objectification as dependent on the particular framework of mental

[10] Intentionality as a concept should not be confused with the notion of 'intention' which is used in the sense of 'deliberate act', or *intending* to do something. Intention in this second sense is one of the mental states that come under the broad umbrella of intentionality, and it is only in that sense that the two are connected.

states of the person who is engaged in the this process. This engagement *involves* the objects in a particular way within the intentionality network of that person. (More precisely, this involvement is not of the objects themselves, but of the mental *content* within which these objects are embedded.) The 'same' object can therefore be invoked differently by different persons, and so embedded in different mental contents, so as to provide different phenomena. When the phenomenological method insists that "objectification is always an active achievement of the ego" (to quote Husserl once again), this achievement is realised through a number of intentional acts, dependent on the mental states held within the framework of intentionality of the individual person. It is through this process that immediate experiences or phenomena can be understood. The strict phenomenological definition of a phenomenon is accordingly simply this: "Any object that is viewed from a *particular point of view* constitutes a phenomenon." This *particular point of view* is provided by the intentional acts originating within the intentionality network. Again, we have to repeat that the phenomenological method is primarily a *descriptive* method, being one that attempts to describe the immediate experiences of creating phenomena in the mind, which are pre-linguistic and pre-predicative- by reducing all statements about phenomena to statements about intentional acts. Without the involvement of intentionality, it becomes impossible to talk about phenomena.

The importance of intentionality can be seen primarily as residing in its *external significance*, in the sense that it links internal mental states with the outer world in terms of what that particular mental state is *about*. However, this stance is not limited to dealing solely with objects in the external world. In fact, if intentionality conditions the creation of phenomena, which constitute our innermost notions of objects, and because our mental states can be directed towards many other possible objects, (imaginary, constructed and even commonly non-existent) this framework appears as one which certainly encapsulates a very broad range of human judgements. This framework is then the most suitably supported by a general theory of objects that includes all possible kinds of objects. Such a theory was provided by Meinong (see Findley, 1995), already in the late nineteenth century and during the beginning of the twentieth century, but because of a general shift in view it nowadays re-emerges and is being re-discovered in many different fields, and fields that are not so obviously connected to philosophy. Abbott (2000b) gives a short introduction to this theory and why it may become relevant to hydroinformatics as well.

Meinongian theory is extensive and multifaceted, and is mostly known for its provision of an extensive taxonomy of those things which Meinong called *Gegenstände,* a word which is usually translated by the familiar word 'objects'. It includes existing objects in the external world, constructed objects in our inner world, and even non-existent, including impossible, objects (described as *subsistent* by Meinong). Obviously this extensive taxonomy is compliant with the broadest definition of an 'object', as 'anything towards which we can direct our thoughts', and so anything that can subsequently become a subject of predication. In fact the Meinongian theory of objects is so general that it even includes judgements, assumptions, or propositions, as special kinds of 'objects', which Meinong called *Objecten,* translated as 'objectives'. The most significant contribution of Meinong, however is that towards the end of his life he provided a *unified theory of objects and theory of values.* This aspect is also the most relevant for our immediate concern with the judgement-forming processes within NDDSSs. Meinong himself first developed these two theories independently, and only in his later works he discovered that these two theories are compatible and in fact unified. Abbott (2000b), explains the relevance of such unified theory for sociotechnology in general:

> Sociotechnical processes are largely driven by gradients in the values of the objectifying agent that are in turn associated with both this agent personally *and* the resulting participating objects. The unified theory then makes possible an analysis of the motivations (based on intentions) of persons relative to their artefacts and, similarly, an analysis of the properties of the artefacts relative to the motivations/intentions of the participating individuals and social groups.

When dealing with judgements of value, and trying to understand the gradients of value, we need to have a unified analysis of both the participating objects *and* the intentionality networks of the objectifying agents. Only through such an analysis can the processes of forming judgements of value be understood. The need to work with such an all-inclusive theory of objects as provided by Meinong is nowhere as clear as in the realm of valuation. In this realm the *intentionality network of the agent involved in valuation in fact conditions the existence of objects*. As Findley expresses it (1995, p.268-269):

> Much that we value profoundly lies quite beyond the range of causal action: we value things that have existed, or that will exist, or that are present in remote parts of the world: we value absences and differences as much as what presently agitates and pushes us. The existence relevant to valuation is an existence for thought and belief, which may not be a real existence at all: it is , in terms of Meinongian psychology, an existence mediated by *judgement*. Value feelings are therefore as much judgement feelings as they are existence feelings.
> And with its connection with judgements, and with the existence presented by judgements, valuation achieves something of the objectivity of the latter.

Even if this framework raises our level of understanding of judgements as means for achieving valuations, the very nature of this process limits the possibilities for its formalisation. Because the origin of all judgements is in fact within the realm of *experience*, all we can do is try to provide some descriptive representation of this process. Such a representation of the whole processes of decision making, where the judgements play central roles, is proposed by Abbott (1999a, 2000a; see also Yan et al., 1999):

$$(\text{beliefs}^{11}, (\text{facts}(\text{data}))) \rightarrow \text{attitudes} \rightarrow \text{positions} \rightarrow \text{judgements} \rightarrow \text{decisions} \rightarrow \text{actions} \qquad (1)$$

The form of this representation is borrowed from category theory in the sense that the links that are connecting these objects are regarded as representing *mappings*. This is of course a very simplified representation, and the steps in the decision making process always involve many feedback loops, and specifically data may intervene in many other places and not only in its interaction with beliefs. This representation may however be used to express something of the structure of this process, even as it should remain clear that these processes with all

[11] Belief is here introduced as a basic mental state, and this does not mean that other mental states are excluded.

their feedback loops are only being realised through *experience*. The design of judgement engine components of NDDSSs therefore cannot use any of these descriptions and representations as prescriptive guidelines. Once the judgement processes are so understood, the actual attention in the design of judgement engines has to be turned towards designing environments where such *virtual experiences* can be created.

What can we then discern from these theoretical considerations in terms of the design of judgement engines? The basic view that is advanced here is that the judgement-forming process is equally dependent on the objects and the mental model of the judging agent. It is a view that is sometimes named *interactionist* (see Klinkenberg, 1996, p.100), and may be contrasted with the traditional dichotomy of the subjective/objective view. The actual mode of interaction is governed by what we commonly call experience. Even though we try to use formalised descriptions of this process in order better to understand it, the actual design of the judgement engines can only benefit if we turn to the question of *how* this interaction actually occurs. The answer to this question must be sought in the study of processes of *signification and communication*, as the embodiments and transmissions of all experience. General semiotics is the field that studies these processes at a level that can provide many guidelines for the design of judgement engines. The provision to the participants in the NDDSS with such virtual experiences that will enable them to make autonomous judgements can only be achieved by the appropriate use of *signs and sign vehicles.*

This field is relatively new to hydroinformatics, and has not been studied sufficiently. In the design of modelling systems orientated predominantly towards specialists in water and environment, the need for a systematic study of the design of user interfaces has not presented such a crucial requirement. The discourse of this user base has been relatively limited and there have not been serious misunderstandings in this process (although we are capable of still distinguishing 'good' from 'bad' user interfaces). The design of judgement engines for NDDSSs will not permit the continuation of the same attitude. The broad discourse of an NDDSS requires a much more systematic approach to the designing of judgement engines and the theory that can offer such an approach is the theory of semiotics. If the first step in such a design is the customised and personalised provision of the facts coming from the knowledge centre, this customised 'content' can enable the judgement–through-experience forming process only if it is delivered with a proper choice and arrangement of the sign vehicles involved.

Understanding the central role of values, which are depending on the users' belief system, the next requirement for the judgement engine is to allow the users to put value judgements on the proposed courses of action *autonomously* and with a proper respect for the data relevant to their interests. The key guideline is *not* to impose any established system of valuations (and thus *not* the one of the designer of the system or the one of the most powerful interest involved in the process, or even the one with the strongest valuational consensus) as a norm or standard for all other participants in the process. The judgement engines have to be designed in such a way that the participants are free to provide their own value judgements. A valuation system that may be representative for a particular community involved in the decision making process may *emerge* from the participants' interactions, but the design of judgement engines should not inherently assume any such system. The design of judgement engines through employment of semiotic structures should be especially careful about this aspect. While a certain structure of meanings will inevitably be presented to all users by the sign vehicles chosen by the designers of the system, a signification structure that would steer the valuation process in one or the other direction may easily have disastrous consequences. It may in fact convert an NDDSS into something that works in the diametrically opposite direction to that which was originally intended. Such strategies have already been implemented in our hypermodern condition, and the particular use of these semiotic structures as 'technologies of persuasion' are well known phenomena in the contemporary media (see Thesis 5 of Chapter 2). Although the study of these technologies is indispensable for the proper design of judgement engines, the always present tendencies of abusing NDDSS through their employment are calling for extreme caution and vigilence, as already emphasised by literature on postmodernism generally.

On the other hand, in any NDDSS there will be some issues for which the valuation is relatively straightforward, and even uniform across all the participants in the process. This is especially the case for those issues that are easily represented in the form of quantitative information. If, for example, one of the issues about which a decision is to be taken is the price that has to be paid for water, it is obvious that almost all the participants will prefer alternatives with lower outcomes for a given water price. However, the problems arise when there are large number of issues involved, and especially when many of them cannot be represented as quantitative information. If we look at the approach taken in various environmental impact assessment procedures and tools, the value judgements for such issues are usually encapsulated in so-called 'value-functions', which are relationships between certain facts and user-perceived 'environmental qualities' or 'values'. Each 'environmental quality" or 'value' is usually expressed on a scale between 0 (very bad) to 1 (very good). These relationships are usually pre-prepared by experts and the user only implements them in the assessment procedure. An example of such value functions used in one such system (MIKE Impact, DHI, 1997) is presented in figure 3.5:

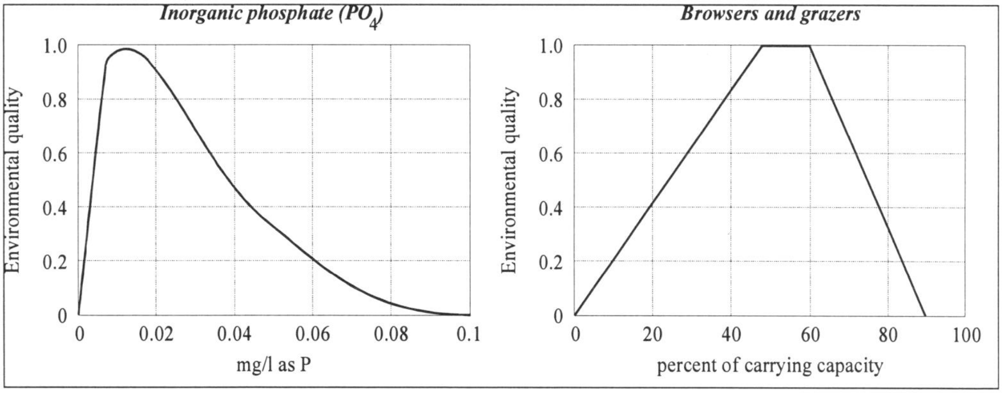

Figure 3.5. Examples of value functions

This kind of expert knowledge should be provided in the judgement engine components of NDDSSs as well, but in a more appropriate form and mainly as information needed to support the judgement-forming process. The temptation to 'lock' the judgements in the value functions should however be avoided. Many participants will have specific knowledge about their local conditions, or have a different reasoning altogether as compared to the one underlying this expert advice, and this may lead to different patterns of valuation. In fact the judgement engines of NDDSSs will have to allow for judgements which are in a sense 'looser' as compared to the structured ones provided by traditional tools and systems for impact assessment, which are based on formalised procedures. Frequently, the only meaningful judgement provided by a given user about a certain state of affairs may be in simple linguistic terms, such as 'good' or 'bad'. This indicates the potential for employing fuzzy sets for conversion of such linguistic terms into fuzzy numbers that can be used for a further processing of the provided judgements.

Such further processing of the provided judgements will be necessary within an NDDSS for two main reasons. Firstly, it may be expected that an individual user/participant of the system will provide judgements about several issues of his or her concern. This means that these judgements will have to be aggregated in such a way that an overall position of that user towards the proposed courses of action will be presented both to that user and to all other interested users. Secondly, in order to facilitate the processes of negotiation and collaboration that should follow the judgement forming processes, the judgements and evaluations of all users need to be aggregated. This aggregation should somehow provide the overall response of the whole community involved in the decision making process (the so-called 'social landscape'). For supporting these tasks, an obvious choice is to use the already well-developed methods of multi-criteria analysis. The same methods that have been used to support the decision making processes within the existing decision-making paradigm, where they were envisaged as providing 'ultimate solutions', can now be turned around and employed to support processes of collaboration and negotiation.

The final requirement of the judgement engine component is again the provision of transparency during the judgement-forming process. Following the representation of the decision making process given in (1), we may address the transparency of the process from a similar, but now reversed representational structure:

$$\text{actions} \rightarrow \text{decisions} \rightarrow \text{judgements} \rightarrow \text{positions} \rightarrow \text{attitudes} \rightarrow (\text{beliefs}, (\text{facts}(\text{data})) \qquad (2)$$

The actual judgement-forming process is not at all transparent by itself when presented in the form of this mapping. The transparency of the judgement engines must be therefore understood in the sense that data and actions in this structure act as primary observables. It is the transparency of these two components that has to be provided in order for the user to be able to deduce the judgements, attitudes and beliefs of all the other participants in the process. In some cases the judgements themselves can be made explicit and they may also act as observables. It is only by supporting this reverse process as well that the judgement engines will be able to mobilise to its full extent the scientific and narrative knowledge, and to support the processes of secondary reflection among the users/participants of the system that may lead towards positive changes in their attitudes (Thesis 4 of Chapter 2 again).

3.4.3 The platform for collaboration and negotiation

The platform for collaboration and negotiation is a component of the total system which will partly be located at the user nodes and partly in the knowledge centre of an NDDSS. The main goal of this component will be to promote collaborative and cooperative behaviour among all participants in the decision making process through their active engagement in a negotiation process. The questions related to the ways in which such behaviour can be stimulated and promoted have been studied extensively in many different fields, varying from sociology, through political science and economics, to decision and management science. Recently these questions have re-emerged in different forms in studies of complex, autopoietic and multi-agent systems. They have now even penetrated the field of general software engineering, and then especially in its efforts towards realising automated cooperation and negotiation of various software components operating in complex systems in order to achieve an effective use of computational resources. In real-world problems, however, the processes of collaboration, cooperation and negotiation still remain essentially *human* activities. Similarly to the formation of judgements, these processes are not amenable to formalisation and it cannot be expected that this component of an NDDSS can be designed as an encapsulation of a given set of procedures which then have simply to be followed. We have to do here with a non-linear discourse, which is not even representable as a linear discourse.

As we have already mentioned, the basis for initiation of the processes of collaboration and negotiation is the aggregation of the judgements and evaluations performed by the participants in the system. This task should allow all the individual users of the system to see their respective individual positions against the background of the overall response of the whole community engaged in the process. The aggregation procedures may vary in different decision contexts and in the design of different NDDSSs. Such aggregation procedures have already been developed in the field of 'group decision making' (e.g. Hwang and Lin, 1987). In this field, traditionally, the main requirement in the development of such aggregation procedures and methods has been to provide one or another kind of mapping from all the individual preferences into a 'social' or 'group' preference in such a way that this procedure will be perceived as 'just' or 'good'. This requirement, which may attain to the status of an imperative, is quite understandable in view of the final goal of all these methods, namely to provide formalised procedures for group decision making that can be implemented for a wide range of group-decision problems.

This formulation of the group-decision problem has been central to the systematic study of so-called 'theories of social choice', and the mapping from individual preferences into social preference has been frequently described through the social choice functions (mainly aggregation procedures based on preference voting systems), or social welfare functions (aggregation procedures based on individual preference orderings). One of the most influential of contributions in this area has been that of Arrow, as initiated already in the 1950s, in which Arrow proposed a set of apparently reasonable conditions that such an aggregation procedure must satisfy, and consequently proved the *impossibility* of designing and implementing such a procedure without relaxing some of these same conditions. This result is nowadays known as Arrow *impossibility theorem*, and for researchers in economics and theory of social choice Arrow's first and central result is comparable to Gödel's celebrated theorem in logic about the impossibility of obtaining certain proofs of inner consistency in formal systems. Many researchers after Arrow have tried to challenge this result on different grounds, and even argued that Arrow's conditions are too restrictive (pointing towards the need to remove some of Arrow's conditions), and are in fact obstacle to the reasonable formalisation of an ethical 'social choice'. The research subsequently progressed to such an extent that nowadays for each of Arrow's conditions there are impossibility theorems that *do not employ* that condition. (For a thorough analysis of this topic up to its date of publication, see Kelly, 1978)

We mention this development here just as an indication of the limitations that have already been identified in aggregation procedures that *rationalise* the process of ethical social choice on the basis of choices of individuals. As we have already discussed in Chapter 2, many communities do not employ such rationalised procedures at all for their social choices, and now we observe that the results of Arrow and his successors point towards the inherent limitations in the rationalised formal approach which has been proclaimed as dominant over all the others during the modern era.

The aggregations that will be implemented in the NDDSSs will most likely follow one of the approaches indicated above, but then with a full awareness that the outcomes of these aggregations cannot be offered as final solutions. Instead, they must be considered as *starting points* for subsequent processes of collaboration and negotiations. Many methods proposed in the area of group decision making, such as the Delphic methods (see again Hwang and Lin, 1987) have proposed that the decision making process should be an *iterative* process, but so far this has been exploited only within such purely social domains as are regarded as the traditional locus of actual decision making. (For examples of this approach in the field of environmental impact assessment, see for example Stauth et al, 1993; Webler et al, 1995).

Within an NDDSS this iterative approach is central and the interactions among the participants is envisaged as bringing new and emerging decisions, which perhaps could not be foreseen. This process must proceed on the assumption that the initial aggregation will have to be continuously updated and modified in accordance with the 'moves' made by the participants in the system/game (as follows from Thesis 6 of Chapter 2). The proper location for the components that will be responsible for these updating tasks is obviously in the knowledge centre.

On the other hand, if the aggregation is the starting point for the processes of collaboration and negotiation, there must be components at the user nodes that will present the aggregation results to all participants and relate these to the own participant's positions. Furthermore, these components have to enable the presentations of the same information about *other* participants in the process. Existing corporate interests, for example, or established common interests, need to be presented to the individual participants in such a way that these participants will have a clear picture about their own position at any moment during the process or play. At the same time, this picture needs to present the dynamics of change, relevant to that participant, occurring elsewhere in the decision making domain. The design of these components, similarly to the judgement engines, will have to employ sophisticated arrangements of signs and sign vehicles for presenting this information. The term 'social landscape' that we have used for describing the aggregate information that needs to be presented to the users, already implies that one possible approach towards the presentation of this information is through dynamic maps (again using GIS facilities). Developments in this direction have already been reported in Tuthill et al. (2000), and especially in Jankowski (2000). In any case, these components of the platform will need to convey knowledge about various kinds of *closeness* between a given individual user and other groups or individuals participating in the process. This may be physical closeness, but it also includes closeness in interests of concern, or judgmental and behavioural closeness. Information of this kind may be then used for formulating definitions of certain *distances* between the positions of various participants that may be utilised as a measure for guiding the subsequent cooperation and negotiation processes.

Once the environment is prepared in the manner described above, the actual negotiation and collaboration processes will proceed towards the ultimate goal - reaching complying decisions. Before turning to the processes themselves, we need to identify the modes in which such compliance with some proposed courses of action can be realised. In most of our present societies, when dealing with conflicts and resolutions of conflicts through negotiation processes, the attainment of a 'resolution agreement' has been understood in two closely related but still quite essentially different notions. The first one is the so–called 'fair agreement', a notion predominantly associated with an 'Anglo-Saxon' approach to conflict resolution. The basic thinking behind this approach is the perception of negotiation parties as *individual* parties, who seek to reach an agreement that will satisfy their interests in a manner that is just and fair, without *imposing* solutions. The criteria of *fairness* are of course varying, and they are themselves frequently being developed within the negotiation process. On the other hand, a notion which may be associated with the 'Continental European' way is that of *consensus*. In this approach, the fairness is again central in the negotiation process, but the parties engaged are frequently seen as constituting a *community*. Here, the agreement to be reached needs to satisfy not only the interests of the individual parties, but also to maintain, or even strengthen, the bonds that are holding together this community. This second notion sometimes involves also the notions of acceptance of certain solutions 'for the sake of the community', even if the individual perception may be that these solutions are not completely fair.

These approaches are the most predominant in any multilateral negotiation today, and the attainment of complying decisions is most frequently sought either through arriving at a fair agreement or at reaching a consensus. Of course there are still many communities where different rules apply, and many complying decisions are being reached through different modes. In fact, the first two approaches are modes of a *democratic* approach to the conflict resolution processes, while other, less democratic, modes are nearly always linked to compliance conditioned by submission to a certain authority (which can be of different kinds, such as religious, professional, political, etc.). In reality, there are elements of both these groups of approaches in any negotiation process, but unless the community is highly totalitarian or dictatorial (where the very notions of collaboration and negotiation are becoming meaningless), there will always be a space for attaining better outcomes by engaging in these processes. An NDDSS should be designed in such a way that it will support democratic approaches to negotiation and collaboration processes, and limit as much as possible the opportunities for abusing the process.

In order to understand what kind of negotiation processes an NDDSS can be expected to support, we will follow a distinction which has been recently proposed between the so-called *positional* negotiations and *principle* negotiations (see, for an introduction to this distinction, Fisher and Ury, 1987). This distinction introduces the principle negotiations approach as superior to the positional one, and more promising for achieving benefits for all sides involved in the process. The basic distinction between these two approaches is presented in Table 3.1:

Table 3.1

Positional Negotiations	**Principle negotiations**
Main	*Characteristics*
Competitive	Collaborative
Distributive	Integrative
Win-lose	Win-win
Zero sum	Non zero-sum
"claiming value"	"creating value"
Goal: maximise own interest	Goal: maximise joint outcomes
Dominant	*Strategies*
Manipulation	Cooperation
Forcing and coercion	Sharing information
Withholding Information	Mutual Problem solving

Obviously, the principle negotiations are more favourable, but it should be realised that any negotiation process usually contains elements of both of these approaches. As explained in Chapter 2 (Theses 2, 3 and 4), an NDDSS should support the principle negotiations approach, and discourage the positional approach as much as possible. There are four main conditions for a principle negotiation process, as proposed by Fisher and Ury (1987). If we go through these four conditions we can see how a system like an NDDSS can provide them:

1. *Separate the people from the problem.* This condition is proposed as one that would exclude the involvement of *personalities* in the negotiation process, and focus on the issues instead. Because an NDDSS of the type now proposed will be a computer-based system, where large number of people will be engaged in the negotiation process, possibly without ever meeting each other, it appears that this condition is not problematic. However, the warning about introducing biases through the choice of sign vehicles, already raised when we discussed the judgement engine component, should be kept in mind here as well. The *virtual* representation of personalities can also be damaging to the negotiation process.
2. *Focus on interests not positions.* Even though we have used the word position several times so far, and even discussed the negotiation process in terms of reducing distances between the positions of the various participants, this should not be understood in the same sense as the word 'position' used in a negotiation context. Adopting a positional

approach in a negotiation context implies unwillingness to disclose, or even to search for, the real interests behind any given negotiation position. It is obvious that the transparency requirement in the judgement engine component, and in the platform for negotiation and collaboration as well, is needed in order to overcome such situations.

3. *Invent options for mutual gains*. This requirement is sometimes referred to as 'enlarging the pie'. In a negotiation process, if the focus is on interests, and not on positions, there are usually possibilities to exploit the complexity of issues and interests involved to discover options that were not imagined before the negotiations started. With the large number of persons that are expected to participate in an NDDSS process, this task will need to be supported by components located at the knowledge centre, which will most likely employ some sophisticated data driven methods for knowledge discovery. On the basis of the judgements and evaluations of the participants, these components should provide indications of possible options that might be acceptable to larger groups of participants, and suggest possibilities for building alliances and coalitions. This process may lead sometimes to the building of new 'monolithic parties', such as were not envisaged beforehand.
4. *Insist on objective criteria*. This last condition needs to be understood in the special context of NDDSSs. The 'objectivity' that needs to be supported by the NDDSS primarily refers to the *facts* coming from the fact engine components. The judgement-forming process and the consequent process of collaboration and negotiation together need to provide *emergent* decision making solutions, which means that the criteria on which these solutions will be based will also emerge during the process.

This discussion shows that, although the platform of an NDDSS should support as much as possible the principle negotiation approach, it seems difficult to frame these processes solely following this concept. The processes occurring within an NDDSS are certainly the most difficult to analyse from the negotiation point of view. They involve *many* parties and *many* issues. As Raiffa (1982) explained in his excellent book on negotiations, this is by far the most complex situation to handle analytically. The problems in this case go beyond the powers of methods developed in the field of 'group decision making with multiple criteria', because the parties are able to form coalitions. Raiffa humorously refered to this situation by saying that when studying negotiations we need to count: "one, two, many".

All this suggests that the platform for negotiation and collaboration of an NDDSS will really only provide a *support* for these processes, as processes that will have to be carried out by the participants themselves. Apart from the support in the aggregation, in the presentation of the social landscape, in the exploration of options for mutual gains, and in the support for alliance building, the quality of transparency of the negotiation process needs to be stressed. Without this quality it will be extremely difficult to promote the cooperative behaviour which is central to the success of a system like an NDDSS (and so satisfying the requirements set out in Theses 4, 7 and 8 of Chapter 2).

If we turn to the conditions for initiating and promoting such cooperative behaviour we shall see how the quality of transparency becomes so important. For this purpose we may turn to one extremely simple model that, regardless of its simplicity, captures the essence of this problem. This is the so-called *n-person Prisoners' Dilemma* problem, an extension of a famous two-person game which has been investigated in the seminal book of Axelrod (1984). We will borrow the example presented in Raifa (1982, p346-349). We imagine a group of 101 individuals engaged in a situation in which each individual can choose whether to act cooperatively or non-cooperatively, but such that that the individual personal payoffs will depend on the behaviours of the 'others'. (The number of 101 individuals is chosen for convenience, because now we can analyse the choice of one individual against 100 'others'). If the number of others who act cooperatively increases, individual payoffs for every individual also increases. However, if at any moment an individual decides to act non-cooperatively, his or her personal payoff increases drastically at the expense of small losses to all the others, and these losses are then evenly distributed over the whole group. With some imagined monetary units assigned to these payoffs, the choice of cooperative or non-cooperative behaviour for one individual, against the behaviour of 100 'others', may be presented as in Figure 3.6:

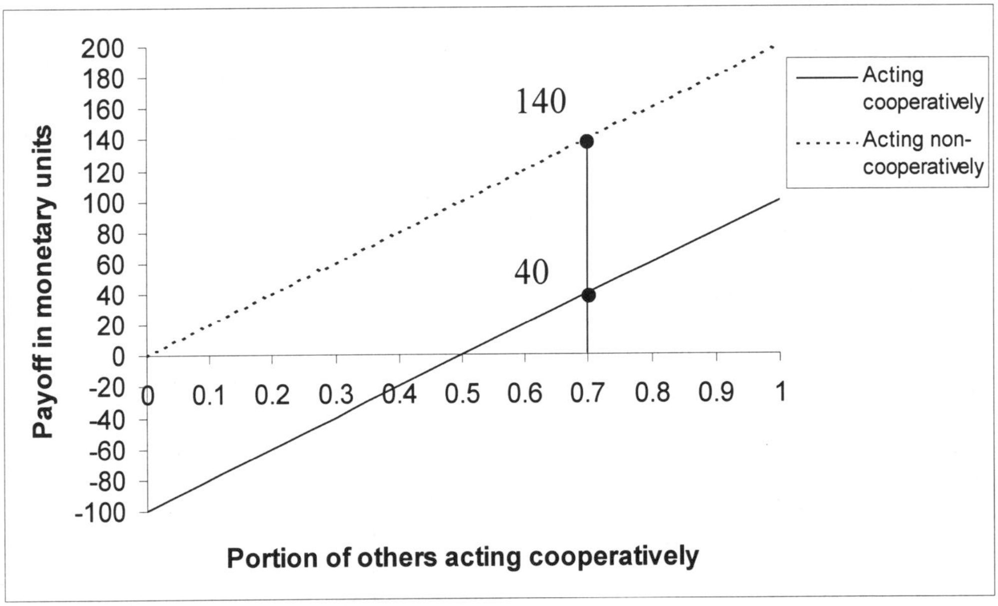

Figure 3.6. Payoffs for an individual participant in a social dilemma game (adapted from Raifa, 1982)

As can be seen from the figure, if all individual participants act non cooperatively, they will all loose. If, for example, 70 of them act cooperatively, the choice for an individual user is ether to act cooperatively and obtain a payoff of 40, or to act non-cooperatively (the so-called 'free rider' behaviour) and obtain 140. The higher the number of those acting cooperatively, the higher the payoffs for the free riders as well. An individual participant is always tempted to act non-cooperatively and to obtain a higher individual payoff, but with each such move, the whole community looses a part of the 'common' or 'social' product. We should note that this picture presents the payoffs for an *individual* participant and not for the community. For the community, the social product increases as more individuals cooperate. For example the total payoff for the community at the same point when 70 act cooperatively, and 30 act as free riders is: 70x40 +30x140=7,000, whereas in the case of all participants acting cooperatively, the whole community gets a payoff of 100x100=10,000[12].

This simple model represents the basic problems related to cooperative behaviour in many real world situations. These are problems that go under different names, such as 'social dilemma', 'problems of the commons', sometimes even 'tragedy of the commons' etc., but the essence remains the same. There is some sort of a 'social product' that can be increased with increased cooperative behaviour. However, increased social product gives more scope to the non-cooperative free-riders. If the number of free riders increases, the social product goes down. In the fields of water and the environment, there are many problems of this kind and it can be imagined that many of the decision making contexts for which NDDSSs will be developed will have this kind of property.

What kinds of approaches can then be taken so as to resolve such problems and how can a more cooperative behaviour be promoted? The essence of the game presented above, as well as of its two-person predecessor, is that *none* of the participants *know* how the others will behave. Many 'social dilemmas' can be resolved by modification of that one crucial rule – allowing the participants to know how the others behave. It is the ignorance, uncertainty and fear about the behaviour of the others that promotes a non-cooperative behaviour. In a setting where the behaviour of all participants is made *transparent*, there are certainly less chances that non-cooperative behaviour will emerge. Within an NDDSS the main role of transparency in the negotiation and collaboration platform is exactly this. In a transparent negotiation and cooperation environment, non-cooperative behaviour is readily identified and easily condemned, while cooperative behaviour can serve as an example that influences others. The achievement of such a state of transparency in the platform for negotiation and collaboration is however a difficult task. From one side, some experiments (see again Raifa, 1982), show that people behind computer screens, tend to behave far less cooperatively than they do in direct human-to-human contact. In our view, much of the desired changes in attitudes can be catalysed by the development of appropriate tools and environments that will offer different kind of *experiences* to the users. The examples of personal and emotional involvement into the virtual worlds of simulacra, especially in the area of entertainment, shows how such environments can be used to stimulate certain behaviour (Thesis 6 of Chapter 2). The design of NDDSS as a whole, and especially the platforms for negotiation and cooperation, must learn many lessons from the successes achieved in these other fields.

[12] These calculations are excluding the one (101st) individual for whose choice the graph is prepared. We may assume that he or she is yet undecided.

Chapter 4

Aquavoice: A Prototype of a Network Distributed Decision Support System (NDDSS)

Fantasy, abandoned by reason, produces impossible monsters;
united with it, she is the mother of the arts and the origin of marvels.
Francisco Goya

4.1 Introduction to ***Aquavoice***

In this chapter we will introduce a prototype of an NDDSS, named *Aquavoice*, which has been developed within the framework of the present study. The design and the development of this prototype have been carried out following the requirements and the general guidelines presented in the Chapter 3, and thus on the basis of the requirements developed in Chapter 2. The main purpose of the *Aquavoice* development is to provide a kind of a demonstrator of the basic ideas presented in this study. It will be obvious, however, that the prototype is still far from a fully developed NDDSS, and it still lacks many important functionalities.

The main idea of *Aquavoice* is to design a prototype of an NDDSS which can be used primarily as a *multi-player role play game* which can be played over the Internet. There are two important reasons for taking this approach. Firstly, this kind of system suits very well to the overall objective of any NDDSS – facilitation of the emergent decision making process through the participation of many interested parties, including both corporate and individual interests. Secondly, the introduction of NDDSS–like systems in our societies may proceed in future exactly through the development of such water- and environment-related Internet based games. This second aspect will be addressed in Chapter 6, when the possible approaches for the introduction of NDDSSs into our present societies will be discussed. As already stated in Theses 5 and 6 of Chapter 2, this approach, which is sometimes called an *edutainment* approach (*edu*cation + enter*tainment*) is probably the best suited to the postmodern condition of our current societies.

The *Aquavoice* prototype has been developed for a hypothetical, but typical case, in which a community is faced with a decision problem related to groundwater abstraction for water supply. One of the most frequent problems associated with projects of this kind is the unwanted lowering of groundwater tables, due to the increased abstraction. In *Aquavoice*, this is the central impact that is of concern to all players/participants. However, the grounds for their concerns are very different between different players/participants, essentially because each of them has a different set of beliefs, interests and intentions. We shall begin this chapter with a general introduction to this particular case. This will describe the decision problem at hand and the nature of the conflict situation resulting from the different roles of the potential players/participants. Before going into the detailed description of the actual design and development of *Aquavoice*, and also in order to present a clear picture of the overall design architecture of *Aquavoice,* we will need to introduce some basic features of Java technology. The most relevant features of Java pertaining to the actual design of *Aquavoice* will be introduced in order also to clarify subsequently the more detailed design of *Aquavoice*. The following section of this chapter will then introduce the design process followed in *Aquavoice* and the methodologies used for the development of its three functional components: the fact engine, the judgement engine and the negotiation and collaboration platform. Then, the description of the actual dynamics of the *Aquavoice* game will follow, with all the steps that can be taken by each respective role player. In concluding this Chapter we shall present some general remarks about the system with some indications of possible directions for its further development. This will hopefully present a clear picture of *Aquavoice*, although we are very much aware that descriptions of tools (and especially a tool like *Aquavoice*) can never attain to the level of intimation with a given tool that can be achieved through the hands-on experience of using that tool.

4.1.1 The hypothetical case of *Aquavoice*

As already mentioned, the central problems in the case employed in *Aquavoice* are those associated with the lowering of groundwater tables due to the installation of new abstraction wells. The actual case is best introduced by following the presentation given in Figure 4.1.

A small town, located at the confluence of two rivers, is planning to implement a new housing development, which requires the provision of additional quantities of water for water supply. The required quantity of water is fixed for design purposes at 6 x 10^6 m^3/year. The only possibility to provide this water is to use the groundwater resources from the aquifer in the vicinity of the town. Previous analyses have resulted in three fixed potential locations where the abstraction wells can be located (well fields 1, 2 and 3: see Figure 4.1). The task of the community is to decide on the appropriate combination of abstractions from these three well fields that will satisfy the required water demand while still being acceptable to the various existing interests. As we shall see, the community in fact faces a decision problem which is a *choice* problem: they will have to choose one of 8 pre-defined scenarios of groundwater abstraction in the configuration of *Aquavoice* considered here.

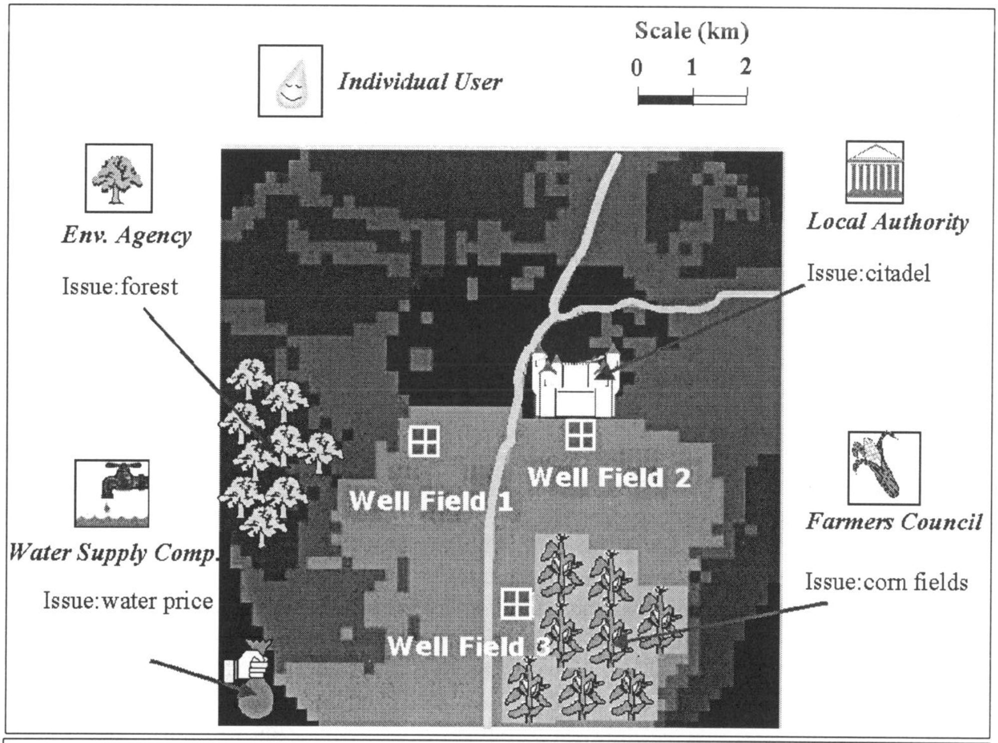

Figure 4.1. *Aquavoice* imaginary case - Introduction

There are two principal kinds of interests existing in *Aquavoice*: the so-called 'institutional' interests, and the individual interests of the concerned citizens. All the institutional interests are representing some sort of 'common' interests of groups of people and are in the present case embodied in four institutions:

-Local authority (represented by the mayor of the town, for example) (LA);
-Environmental agency (EA);
-Private water supply company (WS), and
-Farmers' council (FC).

These four institutions are in one sense major participants in the decision making process: because of their respective responsibilities, they have the power to 'control' four major *issues* which are influenced by the newly proposed project of groundwater abstraction. This 'control', as we shall see later, is exercised by these four institutions through their almost exclusive rights to provide judgements about the influence of the various scenarios of groundwater abstraction on their respective issues of concern. The conflict situation that actually arises in *Aquavoice* will become clearer by presenting the respective roles of these four main institutions.

Local Authority

The Local Authority is very much in favour of implementing the new housing development project. The project is considered as very important for the development of the town because it will attract new businesses and thus create jobs. If the mayor (as representative of the Local Authority) succeeds in realising this project, his or her political position will be significantly strengthened. At the same time, however, the Local Authority is also responsible for safeguarding one particular area of the town, called "Citadel", from any negative influences. It is the area of the historic part of the town, where many old buildings, considered collectively as an important cultural heritage, are situated among parks and city gardens. As presented in Figure 4.1, the Local Authority will be reluctant to accept scenarios of groundwater abstraction which will affect this area (e. g. large abstractions from well field 2). The lowering of groundwater tables could both create land subsidence, which is damaging for the foundations of the old buildings, and have negative impacts on the green areas. The Local Authority has also at its disposal a certain amount of tax-generated revenue. This revenue will probably be most beneficially used by the Local Authority for other activities, which are not related to the current project. However, if the realisation of current project becomes jeopardised because of deadlocks in the negotiations, the Local Authority may decide to use this revenue to support an agreement for this project that is the more favourable to its interests. This support can be offered in two different areas. Firstly, part of the revenue can be used for reversing the negative effects on the Citadel area in case scenarios with large abstractions from well field 2 are inevitable, due to pressure from the other participants. Secondly, part of the revenue may need to be offered to the Farmers Council for paying compensations to the strong farming constituency in case scenarios with large abstractions from well field 3 (in the area of the cornfields) are proposed.

Environmental Agency

The main concern of the Environmental Agency is a forested area southwest from the town which is planted with many indigenous plant species. The forest is also an area that is favoured by many citizens for recreational activities. The Environmental Agency is under the strong influence of some environmental groups and many private citizens to protect this forested area from the negative influences of lowered groundwater tables (primarily due to high abstraction rates from well field 1). An additional important role of the Environmental Agency is related to its responsibility for issuing licenses to the Water Supply Company for groundwater abstraction. In *Aquavoice* the Environmental Agency issues this license for a limited time period, which depends on its judgement about the impact on the environment of the proposed extraction. If the agency judges a certain scenario to be particularly damaging to the environment, the license duration will be very short (for example 5-10 years only), while for scenarios that are perceived less damaging the license duration can be longer (possibly 25-30 years). This responsibility is handed over to the Environmental Agency by the central government, after which the local authority cannot directly intervene (except by persuasion, of course). With this legitimisation, the Environmental Agency is also empowered to influence the choice of the scenarios for groundwater abstraction.

Farmers' Council

The Farmers' Council represents the interests of the farming community, and its main concern are the negative influences of groundwater abstractions on the corn fields located further to the south from the town, close to well field 3. If there are large abstractions from this well field, resulting in a significant lowering of groundwater tables in this agricultural area, the farmers can claim damages from this project and demand compensation from the Local Authority. They will thus accept significant abstractions from well field 3 only under certain compensation agreements.

Water Supply Company

The Water Supply Company is a private company that is interested in the overall situation in its capacity as the local supplier of water, but it remains interested only if it can perform its function profitably. In order to determine the profitability of a given scenario of groundwater abstraction, it has at its disposal a simple economic model. With this model it can calculate the required water prices for certain target profit levels. The economic model itself uses a simple cost-benefit analysis, in which each potential well field brings its own installation and operation costs, which are also dependent on the quantity of abstracted water. One very important parameter that can influence the results of this model is the time period for which the Water Supply Company obtains the license for abstraction (as issued by the Environmental Agency). Naturally the Water Supply Company will favour scenarios with longer licensing periods and with total abstractions obtained from one well field only, because that will yield higher benefits and lower costs, resulting in high profits. Scenarios with short license periods and with abstractions distributed over two or three well fields will yield higher costs and lower financial benefits to this organisation.

With this arrangement, four main issues are established which are more or less accepted as being of common concern to the community:

1) *citadel* (controlled by Local Authority)
2) *forest* (controlled by Environmental Agency)
3) *cornfields* (controlled by the Farmers' Council), and
4) *water price* (controlled by the Water Supply Company)

From the above description of the various responsibilities of the roles involved, it is obvious that the interests of these four institutions are potentially conflicting along different lines. *Aquavoice* is designed to be an environment where, through the processes of their accessing the relevant facts and the forming of their respective judgements by the parties involved, these conflicts will become more transparent, so that possible negotiated agreements for their resolution can be pursued in a more informed and understanding way.

However, the key input, being the one that may significantly influence the outcome of such a procedure will be that emanating from the individual users. The role of the individual user (representing any concerned citizen-stakeholder) is the *fifth* role in *Aquavoice*. A player/participant in *Aquavoice* can take only one of the four 'institutional' roles, but the number of players that can take the role of an individual user is in principle unlimited. The influence of these individual users on the overall process will depend on the number of players that take up this role. In order to clarify the incorporation of the influences of these individual users, we need to present an overview of this role as well, as follows:

Individual Users

In *Aquavoice*, the participation of the individual users in the overall decision making process may proceed in two different modes. Firstly, an individual user may choose to evaluate the proposed scenarios of groundwater abstraction only with respect to the four main issues that are already raised by the four 'institutional' role-players. This is a mode where the individual evaluations may gradually become clustered around the four main institutional evaluations, and the negotiation and collaboration will depend crucially on the behaviour of the four institutional players, with the individual users playing mainly a 'supporting' role. The second mode of individual participation is one where, apart from the four 'default' issues, an individual user can raise an *additional issue* of his of her own concern. Then the individual evaluation of the proposed scenarios of groundwater abstraction will involve *five*, and not four, issues. The new issue raised may be related to a user-given location in the area if required. Clearly, the number of additional issues can also in principle increase indefinitely. One of the key questions in this mode is whether to limit the *kind* of issues that can be raised by the individual users, or to allow for the possibility of raising new kinds of issues, being issues that have not been anticipated in the design phase of the system. This question, as we shall see later, is related to one of the central arguments of the whole approach advocated in this work.

In this introductory section to the particular case around which *Aquavoice* has been constructed, we will not go into details of how the actual game is played. The details about the steps that are taken by each role player and the overall dynamics of the game will be covered in the subsequent sections of this chapter. From this first introduction it should become obvious, however, that *Aquavoice* is designed as a multi-player role play game that can be played over Internet with the participation of a minimum of four players, each of whom takes an 'institutional' role. The total number of players can however be much larger, of course, depending on the number of individual users participating in the game.

The actual design and development of *Aquavoice*, including its deployment on the Internet, was carried out by making extensive use of various Java technologies. Although the design of the different functional components of *Aquavoice* required different conceptual frameworks and methodologies, Java was used both for their actual implementation and for their integration into one coherent system. Therefore, in order to provide a clear presentation of the design of the functional components of *Aquavoice* and their integration, we need to introduce some basic characteristics of Java technology and some of its special features that have been used in specific design and development situations.

4.1.2 Java – The main technology for the development of *Aquavoice*

The explosive growth of the Internet in the last 10-15 years has been associated with a number of different technologies for Internet and World Wide Web (WWW) programming. Traditional programming languages have been extended with additional features for network programming, and a number of new technologies have been developed for the requirements of this new medium. This field is growing so fast that today the number of available solutions for various Internet programming problems is very large. At the same time, new and improved technologies are constantly emerging and replacing the older ones. A useful overview of Internet technologies for problems relevant to hydroinformatics is given in Velickov et al. (1998).

Among the various Internet technologies, Java occupies a particularly remarkable place. This powerful language/platform developed by *Sun Microsystems* has been released approximately over the same period as the WWW, and the whole growth of the Internet and the WWW has been in some way associated with the development of Java. The most powerful feature of Java is certainly the fact that the technology has been designed from the very beginning with network programming in mind. In contrast to many already existing languages that had to undergo some significant updates in order to become usable for network programming, Java had many networking features incorporated in its design already in its first release. Thus, it has been possible for the following versions to include many new features that would easily cope with the constantly growing demands put on Internet and WWW applications. This is a very important characteristic of Java. In separate domains of Internet and WWW programming there exist nowadays technologies that are superior to Java, but there is no other technology that has more integrative power to address the entire complexity of Internet based applications.

Because Java technology itself is such an extensive field we will concentrate here on several of its key characteristics that will enable us to understand how Java works, and on some of its specific features that were used specifically for the development of *Aquavoice*:

1) *Java is both a language and a platform*. This is one of the key characteristics of Java, and lies behind the most frequently repeated advantage of Java – its *platform independence*. The platform independence refers to Java programs being independent of the operating system or the type of computer on which they are running. This is achieved by virtue of the so-called *Java Virtual Machine* (JVM) that needs to be installed on each computer where Java programs are going to run. In fact, the process of building and running a Java program, as presented in Figure 4.2, goes through two phases.

First the source code is *compiled* by a Java compiler into a specific bytecode format that the JVM can understand and the so-called 'class' file is created. Then, the class file can be *interpreted* by a JVM (basically translated into the native machine language of the machine where it is running) that is installed on any other machine regardless of its operating system.

The JVM together with the standard Java Application Programming Interface (API) is referred to as a *Java platform*, and machines and environments that have the Java platform are said to be Java–enabled. The Java platform is nowadays available for all major operating systems, and all major Web-browsers are Java-enabled.

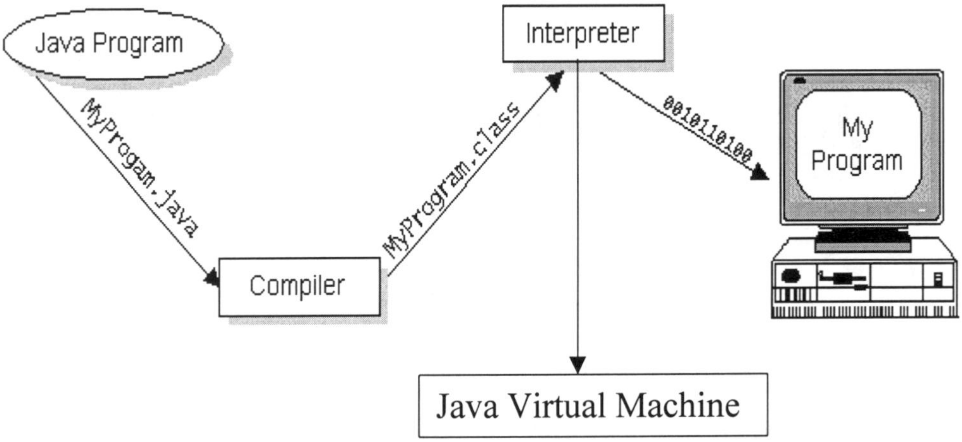

Figure 4.2. Basic operation of a Java program

2) *Java is an object oriented language*. In fact, Java takes the object orientation paradigm further than any other programming language. Everything in Java is treated as an object, including the Java program itself. The transition from object oriented analysis and design into software development is thus facilitated through the design of the language itself. For software developers with their mindset completely situated in the world of object-orientation, Java is easier to learn and to use as compared to many other object-oriented programming languages. By treating everything as an object, the language has also a consistent syntax that is used everywhere in Java programs.

3) *Java is a network programming language*. This is one of the most powerful features of Java. Initially, the popularity of Java was spread because of one particular network programming features – namely the so-called Java applets. The applets are programs that can be compiled to the level of the Java bytecodes (the class files), and embedded into Web pages. When a user downloads the Web-page from a server, the program is downloaded together with the page and the JVM that is integrated in the Web-browser can run the program on the user's (client's) computer. Although Java applets are still very important features of the language, the real power of Java is in its more *general* network programming capabilities. Java has in-built classes (objects) for programming networked applications with a variety of methods, such as:

- *Socket connections*; these are low level connections between two programs that utilise the basic features of the Internet Protocol, such as the Internet Protocol address (IP address) and the port number.
- *Connections with Common Gateway Interface (CGI) programs.* Java applets can easily connect to these standard extensions of Web servers, for sending user specified data to the server, processing the data on the server side, and sending response back to the user.
- *Servlets.* Servlets are special Java programs with similar capabilities to standard CGI programs, but with many additional advantages besides.
- *Remote Method Invocation.* This is a Java-specific solution to the problem of remote code execution by direct calls to the methods of the objects residing on machines different from the local one. A technology which is similar to this one, but designed for such communication of objects written in different languages, is CORBA (Common Object Request Broker Architecture)

Several of these powerful networking capabilities have been used for the design of *Aquavoice*. On the client side, *Aquavoice* makes extensive use of Java applets, while most of extended communication with the Web-server has been realised by using servlets. The servlet programs, as we shall see, are central for the design of *Aquavoice*, because they form the so called 'middle tier' layer between the remote client applications and the back-end server applications. The low-level socket connections have also been used for connections from the servlets to some of these back-end server applications.

4) *Java is a multithreaded language*. The subdividing of a given program into several separate-running tasks, usually called *threads* has been one of the fundamental requirements in computer programming. Whether for making more responsive programs, or for an efficient employment of computing resources, the capability for multithreading has always been important. In Java this capability is built into the language in a transparent manner, in such a way that each thread of an execution is represented by an object. This kind of support is critically important for programming networked applications. In addition to traditional requirements for running multiple threads, in networked applications they are needed for serving multiple client-applications by one or several server programs. The issue of *synchronisation* of multiple threads in these cases becomes crucial, and Java offers significant support in these situations. In the design of *Aquavoice* the multithreading concept has been employed mainly for this purpose.

Apart from those features mentioned here, Java as a programming language has many other unique characteristics. This brief overview cannot possibly cover all of these characteristics, and it is intended only to introduce its basic concepts, and some of the features that are most relevant for the development of *Aquavoice*. At the end, it should be noted that although Java is a full-fledged programming language for building all kinds of software applications, it is nowadays more recognised as a conglomerate of diverse technologies that offer various solutions to the challenges of Internet and Web programming. In addition it should be mentioned that Java is constantly undergoing changes and improvements. This is only partly because as a technology it is still maturing. Most of the changes are in fact demand-driven from the fast growing market for Internet based applications.

4.2 The design of *Aquavoice* and its main components

In this section we will present the design architecture of *Aquavoice*. The design of *Aquavoice* has been determined by considering it both as a prototype of an NDDSS and as a multi-player role play game that can be played over Internet. The main objectives in its design and development were to satisfy as much as possible the design requirements of an NDDSS as presented in Chapter 3, but now, however, by packaging the entire content into a virtual environment with enough entertainment appeal (Theses 5 and 6 of Chapter 2). This second aspect will probably become clearer from the following section (4.3), where we shall concentrate more on the step-by-step description of the actual play carried out by the different role players. The focus of this section will be mainly on the basic approach taken in the overall design of the prototype, and specifically the methodologies applied for the design of the three functional components of *Aquavoice*: the fact engine, the judgement engine and the negotiation and collaboration platform. The following three subsections will be respectively devoted to describing the design and methodologies used for the building of these three functional components.

Following upon the design requirements for an NDDSS, as stated in Chapter 3, and translating these into the specific case of *Aquavoice*, the first objective in the design of the prototype is to provide access to the relevant facts about the various impacts resulting from the proposed scenarios of groundwater abstraction. These facts should be provided to all users/participants in a given *Aquavoice* session in such a way that each should be able to form an autonomous judgement of the proposed scenarios with respect to his or her own interests. This task is achieved in *Aquavoice* by setting up a detailed hydrological model that serves as the fact engine component. The model was developed using the physically-based distributed modelling system MIKE SHE, a product of the Danish Hydraulic Institute. After setting up the model, 8 abstraction scenarios were pre-run, differing the one from the another, by the abstraction distribution over the three potential well fields. These scenarios are then subsequently judged by the participants in an *Aquavoice* session with respect to their own concerns. The model has been deployed on the WWW, so that all the model data are accessible (mainly as maps or time series) through the Web browser.

The judgement engine component of *Aquavoice* was designed following the design requirements for combining the relevant facts with the beliefs, intentions and interests of the users of *Aquavoice*. The basic approach taken in the design of this component was to allow all the users of *Aquavoice* to 'see' the relevant facts and provide their judgements of the proposed scenarios by using linguistic terms, which for the purpose of a subsequent quantitative treatment are converted into fuzzy numbers. The use of pre-defined value-functions has been avoided, assuming that the expert knowledge that is usually contained in these functions can be provided as separate information to the users of *Aquavoice*. Because of their different positions, each of the five types of roles in *Aquavoice* has in fact its own Web-based user interface through which the judgement formation is carried out. The subsequent 'personalised evaluation' of the proposed scenarios by each participant has been enabled by adapting and implementing one of the so-called 'multiple attribute decision making methods'. The actual method selected is common to all the users. The output of the judgement engine component is a personalised ranking of the proposed scenarios for each user/participant. This final ranking, together with all the reasoning behind it, is made transparent to all the users/participants of *Aquavoice* by using dynamically created and updated Web pages.

The negotiation and collaboration platform of *Aquavoice* is relatively simple. The results from the individual rankings of the proposed scenarios are first aggregated, and then the overall ranking of the scenarios representative for the whole community is presented. A crucial aspect in the design of the support for negotiation and collaboration is the automated updating of all the results, both the individual and the overall rankings, as the players/participants make changes in their judgements or evaluations. A specific server-side solution has been implemented for this purpose which 'pushes' the updated results to all the users that are participating in the process whenever a player makes a 'move' (in terms of making a new judgement or evaluation). The actual negotiations among the participants are facilitated by a simple chat-like interface embedded in the interface to the updated results.

The three functional components are brought together in the overall design architecture of *Aquavoice* as presented on Figure 4.3.

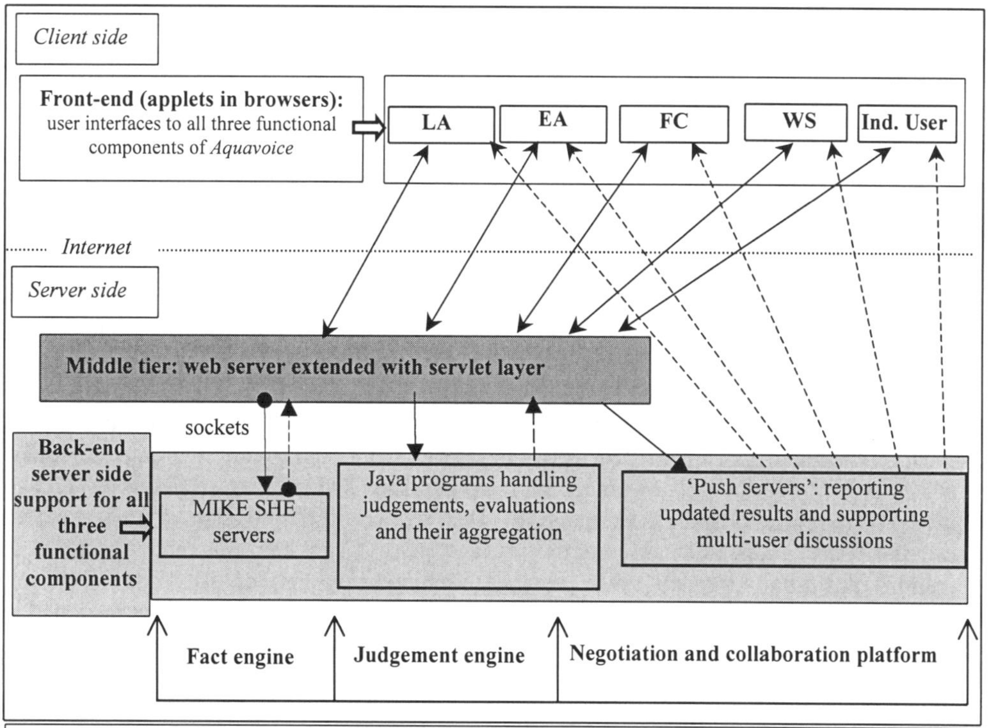

Figure 4.3. Overall design architecture of *Aquavoice*

If we view this figure from left to right, as indicated at the bottom, we can distinguish the three functional components. The view from top-to bottom, on the other hand, represents the design architecture as it was implemented in the development of *Aquavoice*. This view shows that *Aquavoice* as a whole has been designed as a sort of 'three tier' application consisting of front-end, middle tier and back-end part. All three functional components follow this same implementation architecture. On the front end (the client side) the users interact with the system through web pages with embedded applets. Standard HTML, the scripting language JavaScript and Java applets were the technologies used for building these web based user interfaces.

The middle tier of *Aquavoice* is constructed by implementing one particular Java technology, namely the so-called *servlets*. In fact, all the interactivity of *Aquavoice* is achieved through this middle tier layer of servlet programs. Servlets are 'request-response' oriented, server-side Java programs, which are loaded as services on the Web Server so as to increase the server capabilities significantly. As we have already mentioned, they are similar to standard CGI programs, and their primary role is to add interactivity to Web-based user interfaces by creating dynamic responses on the bases of requests coming from the client side. In comparison with CGI, however, servlets have many important advantages. In particular, the applet-servlet communication becomes very efficient, because a servlet can act as a proxy for an applet and connect to servers other than the one that spawned the applet. (Applets are generally limited to communicating only with the server from which they originate). The servlet technology is nowadays frequently used for building 'three-tier' applications, where the servlets provide an intermediate communication layer between the client (usually a Web browser with embedded applets) and various server applications. This approach has also been implemented in *Aquavoice*.

The back-end tier in *Aquavoice* is composed of the various server-side applications, all of them connected to the servlet layer. For the fact engine component, the servlet layer is transmitting the user requests to several separate server programs which run the MIKE SHE executable programs. The communication between the servlets and these servers is through the lower level *socket* connections. As we shall explain later, this design allows the MIKE SHE servers to run on completely separate machines from the one where the Web server is located. The results from the MIKE SHE runs are handed back to the client (the Web Browser) again through the servlet layer. The servlets create the Web pages containing the responses dynamically, or update the applet interfaces with newly requested data. A similar procedure is followed for the judgement engine component, where the user requests are also handled by the servlets. Upon these requests, the servlets run additional Java programs that handle the users' judgements of the proposed scenarios and calculate and update their evaluations. A similar approach is also provided for supporting the processes of collaboration and negotiation. The servlet layer is now connected to several server programs responsible for 'pushing' the updated content to all the users/participants in the *Aquavoice* session, whenever a user makes changes in his or her judgement or evaluation. Because the servlets are 'request-response' orientated programs, using them for the implementation of 'server push' is more cumbersome, and these special back-end server programs handle this task. In fact the updating itself is handled by the servlets and some additional Java programs, and these additional servers are used only for 'pushing' the content back to the clients.

The whole architecture is thus designed to support a genuine multi-user participation, such as is indispensable for *Aquavoice*. The results that are eventually being presented to the users/participants are completely 'event driven' and they depend on the actions taken by all the other players participating in the system. This design enables the 'emergence' of the individual and social preferences for the proposed scenarios as a natural consequence of the direct participation of the users/participants in *Aquavoice*.

As currently implemented, *Aquavoice* is configured to be played by physical individuals: in the absence of such a participant, his or her play is suspended. Future developments are intended to extend the methods presented here so as to allow specific software agents to learn the playing behaviour of each player/participant and to continue to play the game in the absence of the user/player so mimicked. This aspect will be addressed in Chapter 5.

The following three sub-sections will be devoted to the more detailed description of the three functional components of *Aquavoice*. The detailed design of all the three components will always be presented in such a way that it can be referred to the basic overall architecture presented in Figure 4.3. Before going into these detailed description it is useful to take a look at the structure of the experimental Web-site through which *Aquavoice* system can be accessed (Figure 4.4).

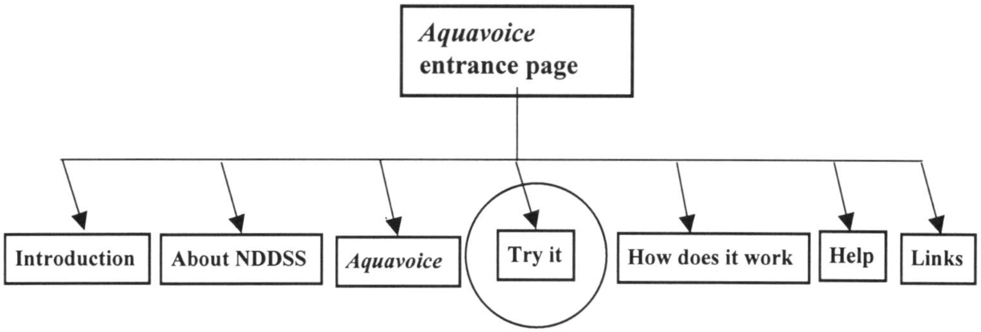

Figure 4.4. Basic structure of the *Aquavoice* Web-site

The site has been deployed on a personal computer linked to the internal network of the International Institute for Infrastructural, Hydraulic and Environmental Engineering (IHE) in Delft, The Netherlands, the institution where this whole study has been carried out. Because *Aquavoice* is still in its experimental phase, at this moment it is still accessible only within the internal IHE network.

The Web-site has a relatively simple basic structure. The entrance page to the *Aquavoice* Web-site contains a menu with links to several other web pages:

- Introduction (basic introduction to NDDSSs)
- About NDDSS (theory of NDDSSs)

- *Aquavoice* (introduction to the imaginary case and the nature of the game)
- Try it (page that leads to actual accessing of *Aquavoice*)
- How does it work (description of an *Aquavoice* implementation)
- Help (link to general help pages)
- Links (page with other links relevant to NDDSSs)

While most of the above pages need further development, the *Aquavoice* prototype is operational and accessible through the 'Try it" link (see Figure 4.5). Once the user accesses this page, he or she can choose to participate in an *Aquavoice* session, or to look at all hydrological data available from the fact engine component – the MIKE SHE model.

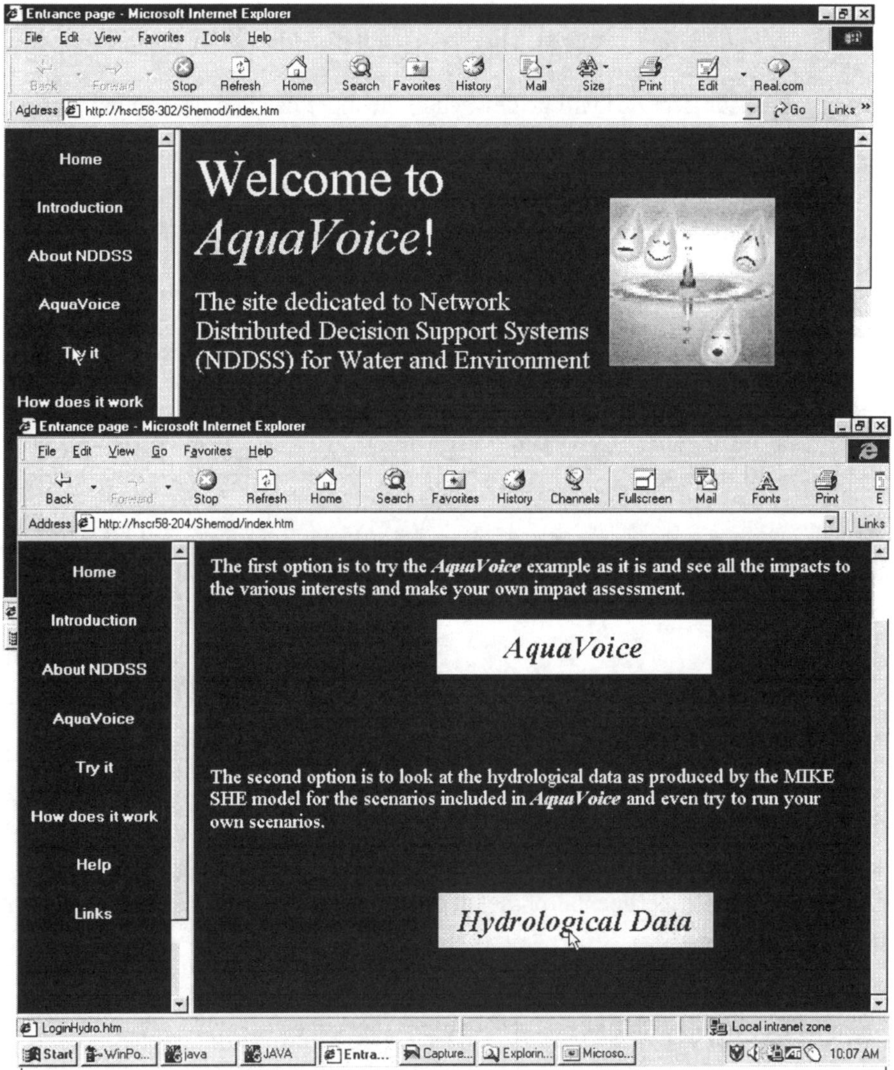

Figure 4.5. The entrance into the *Aquavoice* Web-site

The reason for this separation is in the fact that the actual deployment of an instantiated MIKE SHE model on the Internet has been developed as a separate application, where all the model input and output data available from MIKE SHE are made accessible to all users. When converting it into a fact engine for *Aquavoice*, only selected kinds of model results are actually used during the playing session. By following the link to 'hydrological data', the user will be able to see the complete set of MIKE SHE model data. On the other hand, by following the link to '*Aquavoice*', the user can participate in an actual *Aquavoice* session. The task of building, instantiating and deploying a complete MIKE SHE model on the Internet that can later be used as a fact engine for *Aquavoice* preceded the development of all other components of the prototype, and will be the first issue that will be covered in the detailed description that follows upon this section.

4.2.1 The Fact Engine – the MIKE SHE hydrological model

This subsection will be divided into three parts. In the first part we will give a very brief overview of the MIKE SHE modelling system. The second part will describe the configuration of the actual hydrological model used in the *Aquavoice* case. The deployment of this model on the Internet, for the purpose of converting it into a fact engine component of *Aquavoice,* will be presented in the third part.

4.2.1.1 The MIKE SHE modelling system.

The choice of the MIKE SHE modelling system is directly related to the requirement of using *integrated modelling* approaches for constructing the fact engine components of the NDDSS, as already introduced in Chapter 3. In this respect, MIKE SHE is probably the most integrated commercially available modelling system for hydrological modelling. The development of MIKE SHE is based on the SHE (Systéme Hydrologique Européen) concept developed during the late 1970s and 1980s by a European consortium of three organisations: The Institute of Hydrology (UK), the French consulting company SOGREAH and the Danish Hydraulic Institute (DHI) (see Abbott et. al. 1986 a and b). As a result of this approach, there nowadays exist two SHE versions: MIKE SHE, the commercial package of DHI, and SHETRAN, a modelling package developed by the Water Resources Systems Research Laboratory, at the University of Newcastle.

MIKE SHE is a physically based distributed hydrological modelling system which allows for the simulation of the whole land phase of the hydrological cycle (Refsgaard and Storm, 1995; Storm and Refsgaard, 1996). The basic Module of MIKE SHE is the Water Movement Module (WM) for describing the water movement in the modelled area. On the basis of a model instantiated with the WM module, additional processes can be simulated with a number of add-on modules available for MIKE SHE, such as: advection and dispersion of solutes, geochemical processes, biodegradation, soil erosion, etc. In this overview we will concentrate on the WM module, as this was the only one used in *Aquavoice*. (More information about MIKE SHE and all the available add-on modules is available from DHI, through its Web- site (http://www.dhi.dk)).

The MIKE SHE WM is itself built with a modular structure. Six different process-oriented components for the simulation of the relevant physical processes in individual parts of the entire hydrological cycle are integrated into one computational framework:

- Interception/Evapotranspiaration (ET)
- Overland and Channel flow (OC)
- Unsaturated Zone (UZ)
- Saturated Zone (SZ)
- Snow Melt (SM)
- Exchange between Aquifers and Rivers (EX)

The overall structure of the MIKE SHE, with the identification of the processes that are covered by these modules, is illustrated on Figure 4.6.

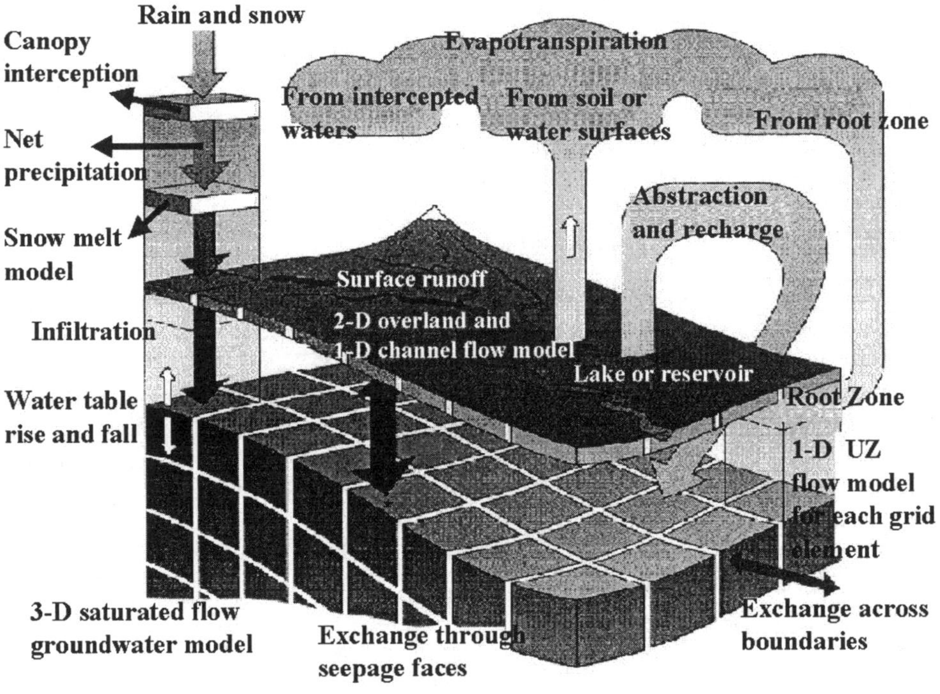

Figure 4.6. Schematic representation of the MIKE SHE model structure (adapted from DHI, 1998)

In MIKE SHE the modelled area is discretised in the horizontal plane by a square grid network. This network is used both by the 2-D overland flow component and the saturated zone (groundwater flow) component; the latter can be modelled with the quasi 3-D (layered aquifer and aquitard approach) or real 3-D approach. Each grid cell is connected from the land surface to the groundwater table with a vertical column in which unsaturated flow is modelled in 1-D. The channel flow (in the river network) is also modelled in 1-D.

Interception and evapotransporation, snow melt, and river-aquifer exchange processes are included in separate components and linked with the other water movement components.

The main equations in the model used to describe the physical processes involved are well known. The channel and the overland flow are described by solving the continuity and the conservation-of-momentum equations (the Saint-Venant equations) in 1-D and in 2-D respectively. For the momentum equation a diffusive wave approximation is used. The stream channel network is assumed to run along the boundaries of the grid squares. (This approach can be changed in the latest MIKE SHE versions, where the channel flow component can be replaced by the MIKE 11 modelling system for simulating the 1-D open channel flows. The version used in *Aquavoice* (MIKE SHE 5.3 for WINDOWS 95/NT) uses the older channel component. The outputs are obtained as water depths and flows in two horizontal directions for overland flow, and in water depths and discharge at the river nodes for the channel flow.

The unsaturated zone is modelled by solving the Richards equation for 1-D vertical flow. The outputs are obtained as soil moisture and unsaturated zone flow in all discretisation nodes in the vertical column above the water table. The Richards equation is solved by making use of two non-linear relationships: the hydraulic conductivity function (K-θ, relation between hydraulic conductivity and soil moisture) and the soil moisture retention curve (ψ-θ, relation between soil moisture and pressure head). These relationships need to be provided for all types of soils that are included in the model.

The saturated zone is modelled by solving the partial differential equation for groundwater flow in 3-D and is applicable both for single-layered aquifers and for multi-layered aquifers. In applications with negligible vertical flow components, the 2-D equation is solved with flow predominantly in the horizontal direction. The outputs are obtained as groundwater heads and flows in time and space.

The modular structure of the MIKE SHE allows for great flexibility in the description and integration of the individual processes that need to be modelled. In some cases, only some of the individual components need to be modelled, such as if only rough estimates of the exchanges from other parts of the hydrological cycle are required. A groundwater model, for example, may be built with approximate recharge estimates, without the full coupling to the unsaturated zone above the groundwater table.

Although this flexibility is useful in some modelling studies, the power and sophistication of MIKE SHE lies exactly in the integration of all these different hydrological processes into one computational framework. This integration is achieved despite of the fact that the nature of different hydrological processes requires their modelling on different time and space scales. The governing partial differential equations that describe these processes are solved numerically with stable and accurate finite difference methods for each individual process component. Different processes are modelled with different time steps, appropriate for the respective processes, and *during* the simulation they are coupled together when their time steps coincide. Special coupling procedures have been designed for this purpose, such as for example the one between the unsaturated and saturated zone.

One additional component, the so-called Frame component is used for managing the proper integration of all different modules during a MIKE SHE WM simulation. This component controls the sequence in which each component is called, and manages the different time steps used for different components. It also controls which components are actually included during the simulation, and whether there is a need to use dummy versions of some components that are irrelevant for the application at hand (and therefore excluded), but only need to provide boundary conditions for the active components. The Frame component is responsible for controlling the data exchange between the components, as well as all the outputs that need to be stored on disk for future post processing. The data flow between the different MIKE SHE components, as ordered within a special integrating framework, is presented in Figure 4.7.

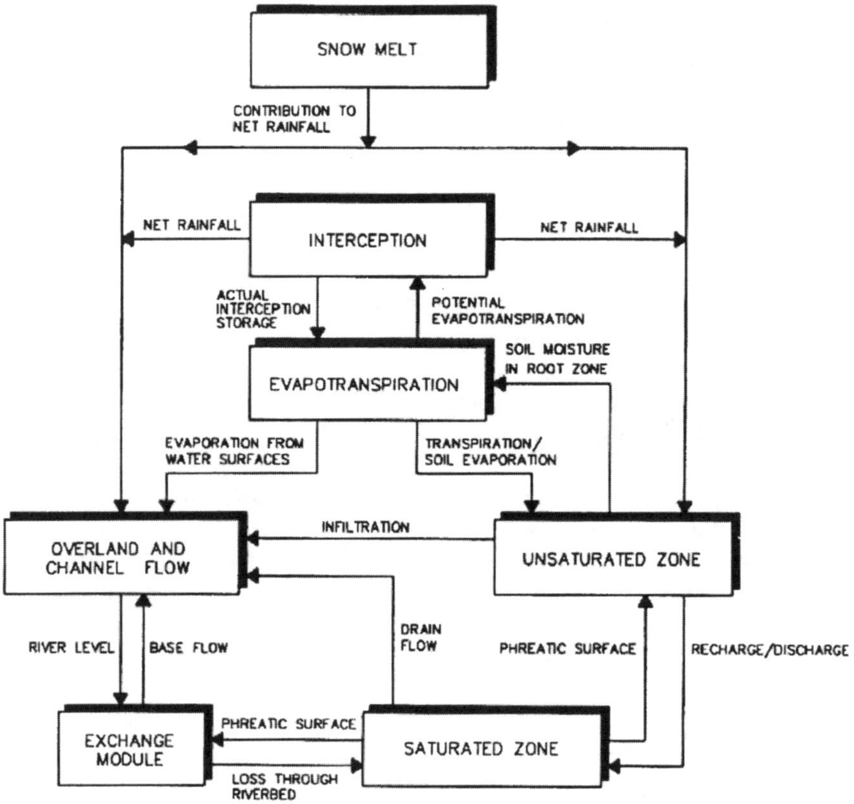

Figure 4.7. Data flow between the different MIKE SHE components (adopted from DHI, 1993)

The MIKE SHE modelling system comes with full pre- and post-processing support. The pre-processor comes with powerful GIS capabilities. Digitization of contour maps, river systems and other spatially distributed information is provided, with interpolation facilities for obtaining point values or grid-cell averages. Graphical editors for working with 2-D data (maps) and river data allow for efficient preparation of all input data for the model. The model outputs are accessible through the post-processing package as maps, cross sections or time series, including animations.

The integration provided by the MIKE SHE modelling system is higher then in any other hydrological modelling system currently available (e.g. Kaiser-Hill, 2001). With the various add-ons available it is a powerful tool for modelling many water and environmental problems. Developments are already under way to extend the integrated modelling capacity even more, by interfacing MIKE SHE with other models that were previously designed for modelling entirely separated water related-problems. For example, DHI is working on interfacing the MIKE SHE with the MOUSE modelling system, used for urban drainage and sewer systems. All these developments show the increasing demand for integrated modelling, which, as we have already mentioned, is crucial for the future NDDSSs.

At the same time it needs to be recognised that the same process of developing integrated modelling system brings new *socio*technical challenges. The achievement of integrating all different components into one computational framework in MIKE SHE is certainly not only technical. Expertise in hydrology and hydraulics was brought together with such field as glaciology, soil physics, meteorology, agriculture and plant development in order to achieve the proper integration. Abbott (1993), described how this 'merging' of knowledge from previously separated domains has been achieved through the design of a *taxonomy of the modelling system* which reflects the structure of the semantic content of the corresponding knowledge involved. Future integrated modelling will require the bringing together of even more separated knowledge domains, which will similarly be enabled by the new taxonomies of the new kinds of knowledge encapsulators.

4.2.1.2 *Aquavoice* model set-up

Because *Aquavoice* uses an imaginary case study, the setting up of the model did not have to follow all the necessary steps in the modelling protocol as required in modelling real world cases (see Refsgaard, 1996). On the other hand, this does not imply that the model used in *Aquavoice* is very much oversimplified. On the contrary, the model is built by inclusion of all the MIKE SHE WM components, and with a complexity which is close to a real world modelling application. The key requirement in the model set-up was to have some reasonable representation of the hydrological processes involved. In order to achieve this, the imaginary conceptual model of the area was combined with some real world data from one small European catchment. The general set-up of the model and all its constitutive components is described as follows:

General model set-up

The area covered by the model is about 100 km^2 (10x10 km). The elevation of the land surface decreases from north to south from about 120 m.a.s.l. to about 70 m.a.s.l., as presented on the topographical map (see Figure 4.8). For the horizontal discretisation a grid of 50x50 square cells is used, each with dimensions of 200x200 m.

Topography **Horizontal grid**

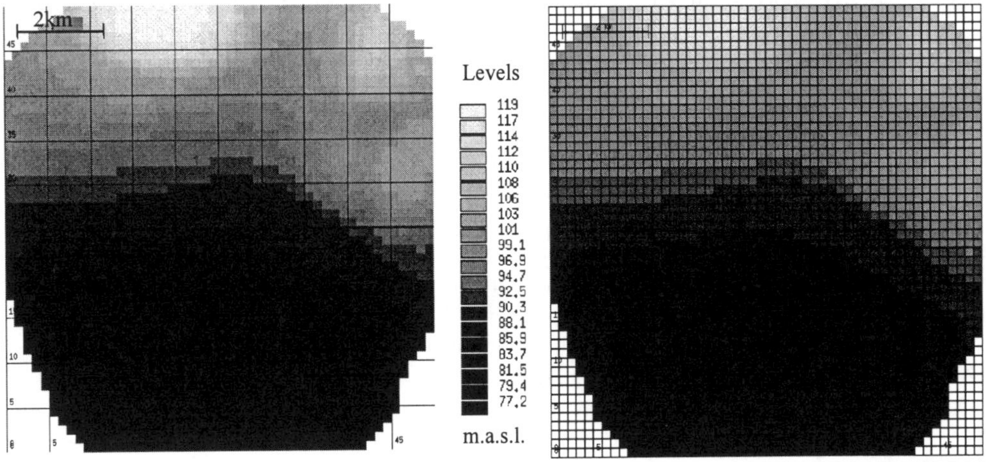

Figure 4.8. Topography and horizontal discretisation of the model

Land use map

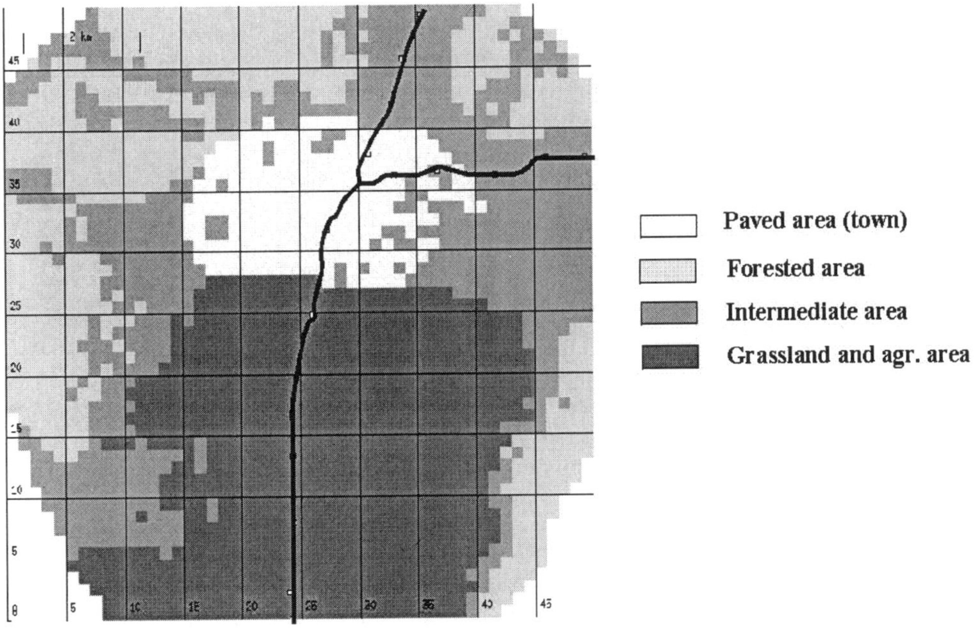

Figure 4.9. Land use map and rivers of the model

On the land use map (Figure 4.9) the location of the town is presented, slightly to the north from the centre of the modelled area. As shown in the map, the model uses 4 different kinds of land use: *paved* area (at the location of the town), *forested* area, *grassland* (including the agricultural area), and *intermediate* areas with mixed land use. The town itself is located at the confluence of two rivers, one larger river flowing from north to south, and one smaller tributary coming from an easterly direction.

Hydrological data for the model

In order to have realistic results from the model, the real data from a small European catchment were taken as actual input data for the model. The main input for the model was the precipitation data. A record of 15 years of daily precipitation data was available, from 1973 to 1991. The average annual precipitation from this available record was 705 mm/year. The simulations in the model were carried out with the daily data for one average hydrological year (1974), with average annual precipitation of 711 mm/year. A uniform spatial precipitation distribution over the whole modelling area was used. Only the simulations from this year were actually used for the *Aquavoice* game. For comparison, and for possible future extensions of *Aquavoice*, simulations were also carried out for three other characteristic hydrological years: 1985 – a dry year with an annual average of 573 mm/year; 1987 – a wet year with an annual average of 826 mm/year, and 1976 – an extremely dry year with only 359 mm/year. The actual model simulations were done using the available daily data, but here, by way of an overview, the monthly averages for the four characteristic years are presented (Figure 4.10).

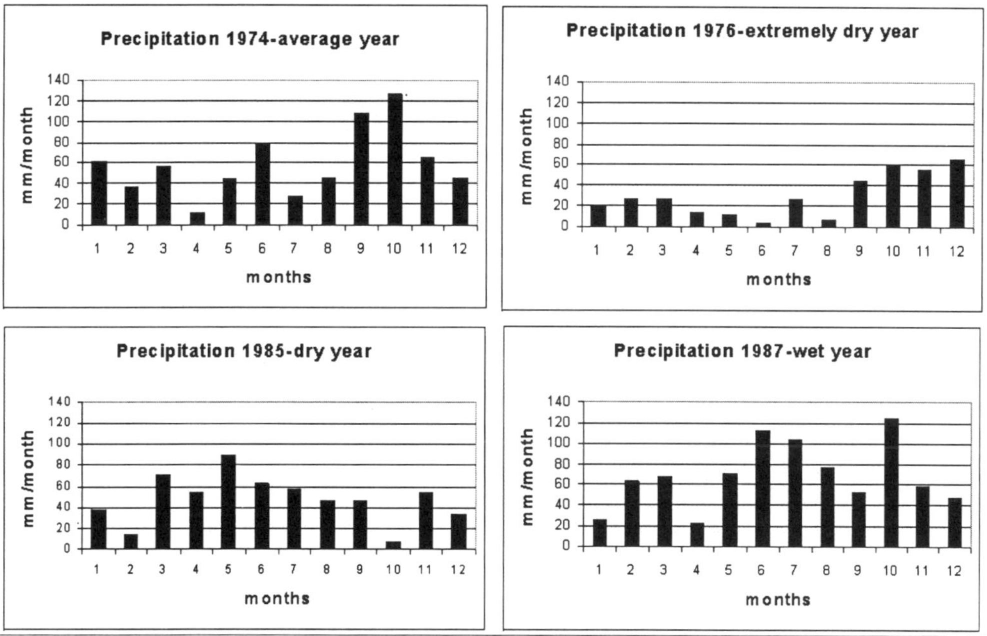

Figure 4.10. Monthly averages of precipitation for the four characteristic hydrological years

Real world data are also used for the potential evapotranspiration, the root zone depth and the leaf area index, for the three respective kinds of land use (excluding the paved area). As shown below, these data are used by the interception/evapotranspiration component of the MIKE SHE WM model. Also, real world data were used for the characteristics of the soils used in the model, which will be discussed under the heading of the unsaturated zone component.

Interception/evapotranspiration component

For its interception/evapotranspiration component, the MIKE SHE WM uses the Danish method of Kristensen and Jensen (1975). The interception is modelled by introducing interception storage, which has to be filled before throughflow to the ground surface takes place. The capacity of this interception storage is dependent on the so-called Leaf Area Index (LAI), which describes the degree of vegetative coverage. For this model, monthly data of average LAI are used for the three vegetation types (land uses).

Actual evapotranspiration is the combination of transpiration and soil evaporation. Transpiration is calculated by using the reference potential evapotranspiration, LAI and Root Distribution Function (RDF). These three parameters are linked to the soil moisture conditions by two empirical functions. RDF depends on the root depth and is calculated assuming a logarithmic variation with depth as the usual distribution of the root mass. Soil evaporation is calculated using empirical functions that describe the dependence of the percentage of potential evapotranspiration that directly evaporates on the actual soil moisture content in the top soil layer. Both transpiration and soil evaporation calculations are done through interaction with the UZ component, since they depend on the actual soil moisture conditions in the root zone. Monthly data for the root depths of the three vegetation types and potential evapotranspiration were used in this model. The method uses several empirical constants, for which recommended default values have been used.

Overland and channel flow component

The overland flow is modelled on the already introduced horizontal grid. Two necessary parameters are specified for the entire model area. The detention storage depth (above which the actual flow occurs) is specified as 1 cm. The Manning number M that describes the bed roughness for the overland flow is specified at 5 $m^{1/3}/s$ (MIKE SHE WM uses the reciprocal of the more standard Manning coefficient n).

The channel flow has in the present illustrative case a very simplified representation: the same cross-section geometry is used at all river nodes, with the same Manning number M of 20 $m^{1/3}/s$. The two upstream boundaries, the northern one of the main river and the eastern one of the tributary, are time series of discharge, while the southern, downstream boundary of the main river is provided as a constant water level. All these data are imaginary, but in order to have some reasonable values for the upstream boundary conditions a much larger regional model was built with similar characteristics to the actual one, which was used as a simple runoff generator. From this model, the discharge time series were read as boundary conditions for the actual model. For an average year the discharge at the boundary for the main river varies between 5.0 and 30.0 m3/s, and for the tributary between 1.5 and 9.0 m3/s. The enumerated river nodes, the cross section used and the specified boundary conditions are presented on Figure 4.11.

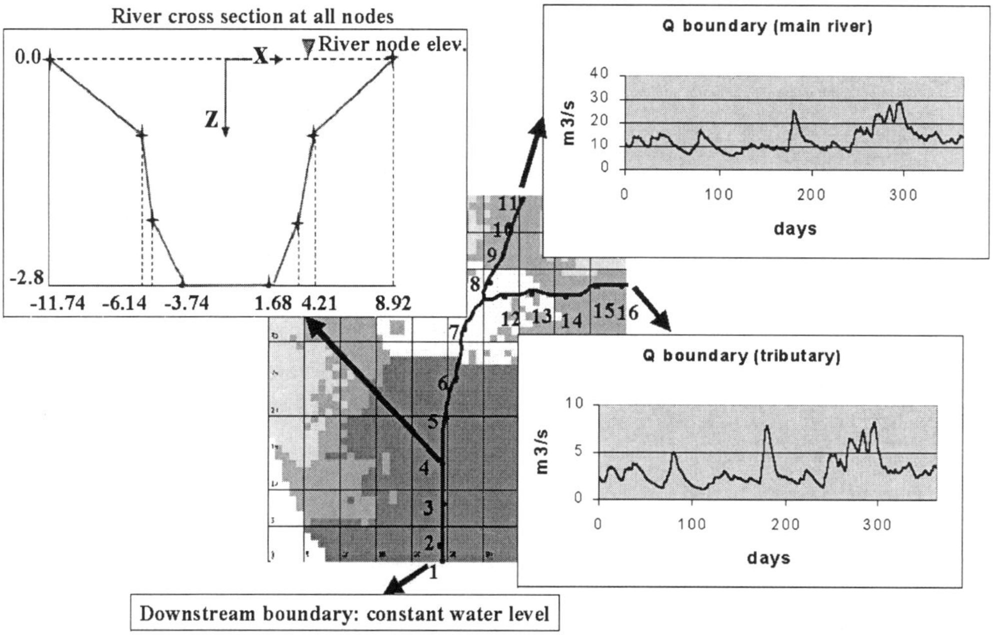

Figure 4.11. Data used for the channel flow component

Unsaturated zone component

For simulation of the flow in the unsaturated zones three different types of soils were used: tilly (clayey) soil, fine sand and coarse sand. These three types of soils are combined into four different soil profiles covering different areas of the model, numbered from 1 to 4. Each profile has the tilly soil on top, but then with varying depth. Deeper strata of the soil profiles are constituted of fine and/or coarse sand layers, again with different depths for different profiles. The spatial distribution of the soil profiles and an example soil profile (for soil profile 3) are presented in Figure 4.12.

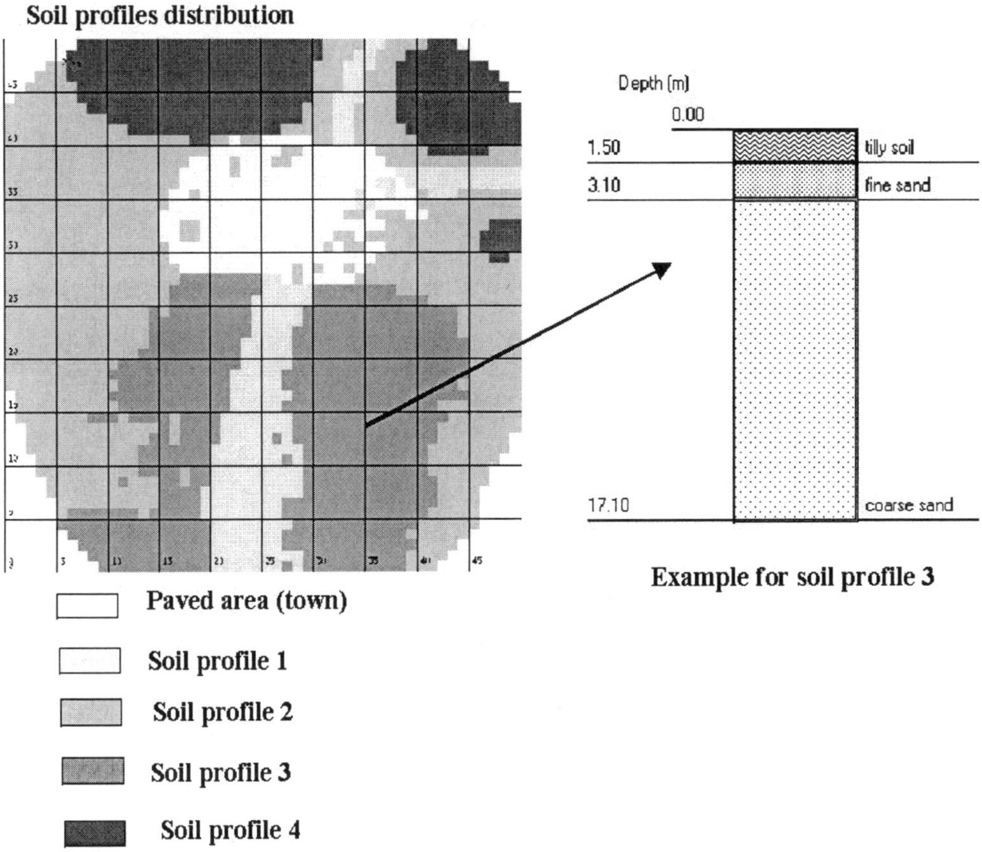

Figure 4.12. Distribution of the soil profiles and an example for soil profile 3

A soil profile is specified at each computational grid of the horizontal network of grid cells. The paved area, where the town is located, is excluded from the unsaturated zone calculations: all the excess water from the town is supposed to be drained directly into the river. The calculation of the unsaturated flow is performed by discretising the vertical column with the soil profile. Closer to the soil surface, in the root zone, a finer spatial resolution is used, with computational points at intervals from 2 to 5 cm. The resolution gradually becomes coarser with depth, and at depths close to the groundwater table the computational points are at 20 cm distance. For each soil included in the model the soil moisture retention function (ψ-θ), and the hydraulic conductivity function (K-θ) are specified. In MIKE SHE the K-θ function is described as a function of the effective saturation and the saturated hydraulic conductivity. The ψ-θ function is specified in tabular form. Examples of these two functions for the soil type 'coarse sand' are presented in Figure 4.13.

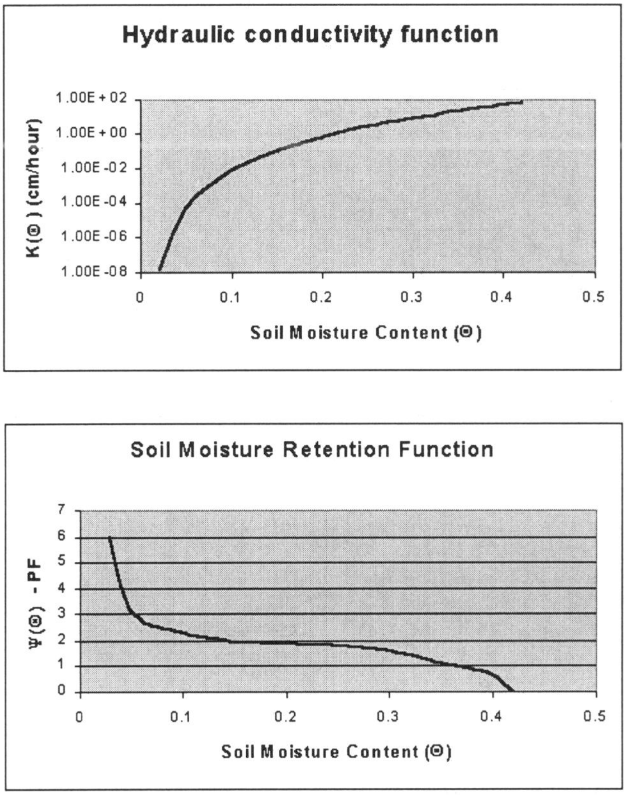

Figure 4.13. Hydraulic Conductivity function and Soil Moisture Retention curve for soil type 'coarse sand'

Calculating the UZ flow in each computational cell of the horizontal grid is expensive and time demanding. In order to reduce the computational time, in this model the optional, so-called lumped, UZ calculation approach was employed. The UZ calculations are then made only for selected grid squares and the boundary conditions from these are transferred to equivalent grid squares elsewhere in the model. The classification of 'equivalent' or 'homogeneous' zones for UZ calculation is determined on the basis of the homogeneity of the soil- and vegetation characteristics and the use of identical boundary conditions. A separate utility for this purpose is provided in the MIKE SHE WM pre-processors that enables this classification to be made on the basis of depth to groundwater table and the available information on soil, vegetation and climate.

Saturated zone component

The flow through the saturated zone is modelled in 2-D in the horizontal direction. There is only one aquifer layer, with a thickness varying between 35 and 55 m. At the bottom of the aquifer an impermeable base is assumed. The spatial distribution of hydraulic conductivity of the aquifer is presented in Figure 4.14.

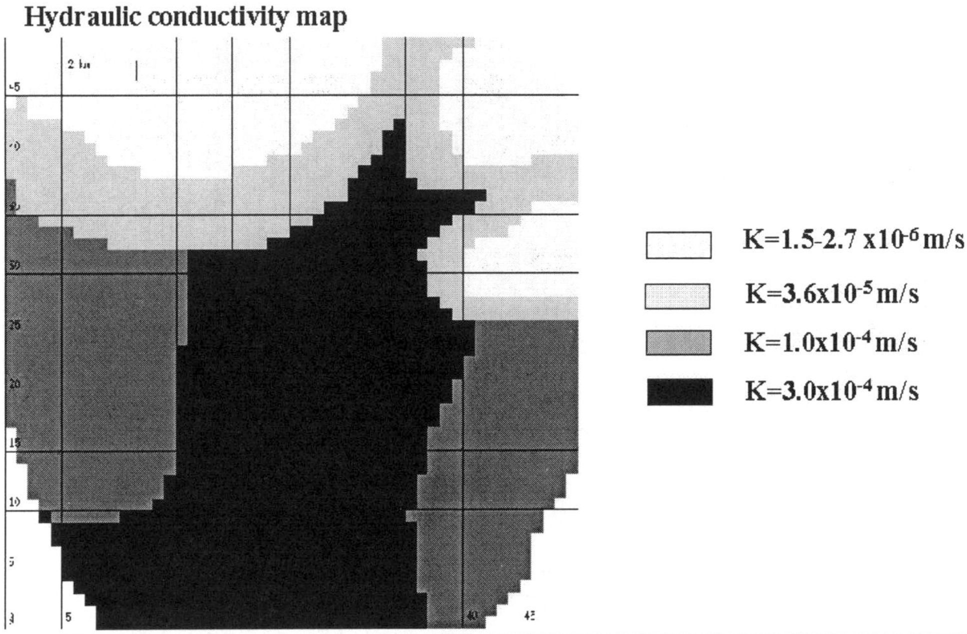

Figure 4.14. Spatial distribution of hydraulic conductivity

The southern boundary is modelled as a constant head boundary. Along all other boundaries constant groundwater flow gradient is specified: this is representative of the natural groundwater flow from northeast to southwest direction.

With this model configuration the so-called 'No-abstraction ' scenario is first simulated. Then 7 scenarios of groundwater abstraction are simulated for each hydrological year. The results from these 8 scenarios are then used by the *Aquavoice* game. Table 4.1 shows how the total amount of water (6×10^6 m^3/year) is distributed over the three proposed well fields in these 8 different scenarios. (For location of the three different well fields, see again Figure 4.1)

Table 4.1

Scenario	Abstraction (m^3/year)		
	Well Field 1	Well Field 2	Well Field 3
Scenario 0	0.0	0.0	0.0
Scenario 1	6.0	0.0	0.0
Scenario 2	0.0	6.0	0.0
Scenario 3	0.0	0.0	6.0
Scenario 4	3.0	3.0	0.0
Scenario 5	0.0	3.0	3.0
Scenario 6	3.0	0.0	3.0
Scenario 7	2.0	2.0	2.0

The results available from the model (daily outputs) for all scenarios are stored on disk in binary result-files. The model contains all MIKE SHE WM components, so that the number of possible results that can be obtained is very large. Just as an indication, one result file of this model stores about 30 MB of data. Accessing the outputs stored in these result files over the Internet, as well as all relevant model inputs, actually becomes the main task of converting the instantiated model into a fact engine for *Aquavoice*.

4.2.1.3 Deployment of the MIKE SHE model on the Internet

The deployment of the MIKE SHE model on the Internet has primarily been carried out within the framework of the *Aquavoice* development. It also serves as a demonstration of one approach for solving the general problem of allowing multi-user access to instantiated models over the Internet. As we have already mentioned, the developed application allows for access to *all* MIKE SHE model results, while only a limited set of these results is used in the *Aquavoice* game.

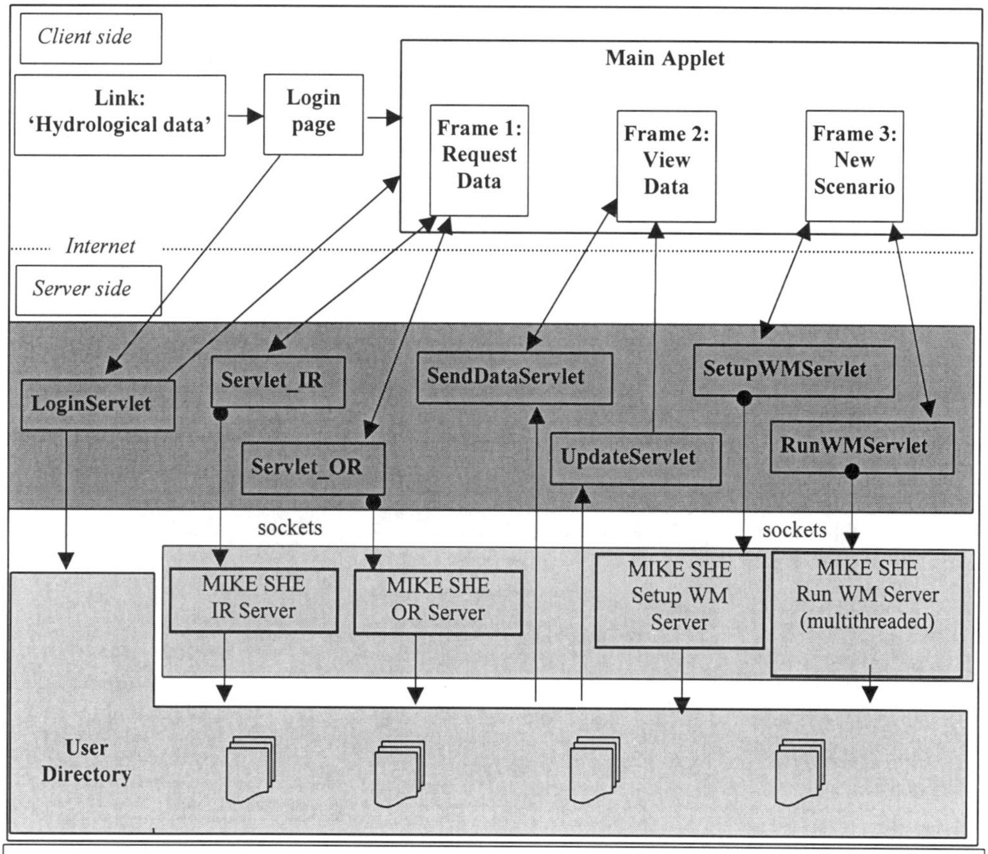

Figure 4.14. Implementation architecture of the MIKE SHE model on the Internet

This application is developed following the general implementation architecture of the *Aquavoice* system as already introduced (see again Figure 4.3). A more detailed presentation of the three-tier architecture used for the fact-engine component is presented in Figure 4.14. This still excludes many smaller supporting programs, but does present the main components of the application.

By following the link indicated by 'hydrological data' from the second entrance page into the system (see again Figure 4.5) the user is led to the application through a login procedure. The user name and password supplied on the login page are used to identify new and existing users, and to create a separate working space on the server for each registered user. This working space contains a specially designed directory structure for the storage of and the access to all data requested by the user. On the server side, the login procedure including the creation of the user's work space, is handled by the *LoginServlet*. The user is then led to the front end of the application.

The front end of the application contains a rather complete Web-based user interface through which the user can formulate his or her request, and then display all the relevant model data. This user interface is completely contained within a Java applet embedded in a Web page. The applet carries three separate frames, each responsible for handling different tasks. The 'Request Data' frame allows the user to formulate his or her request. The steps for formulating a request are presented on Figures 4.15-4.19.

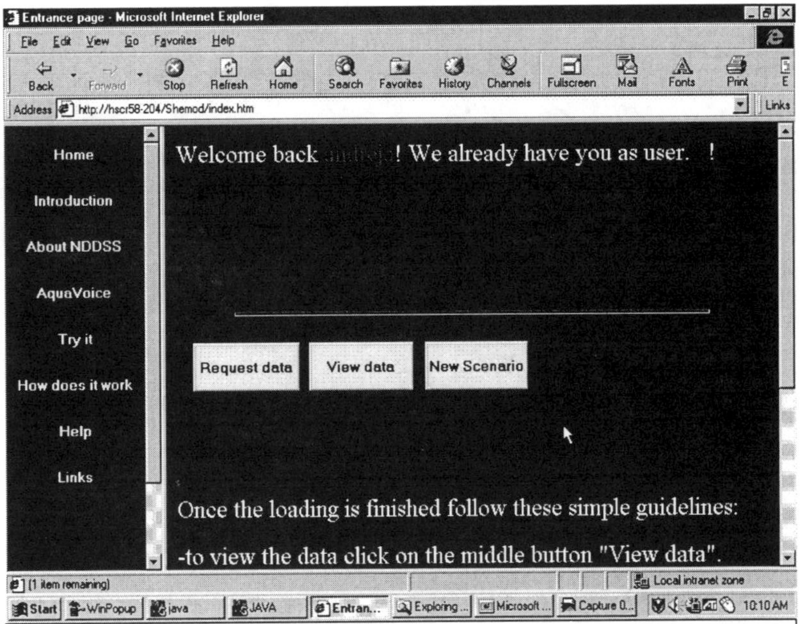

Figure 4.15. Page with the main applet

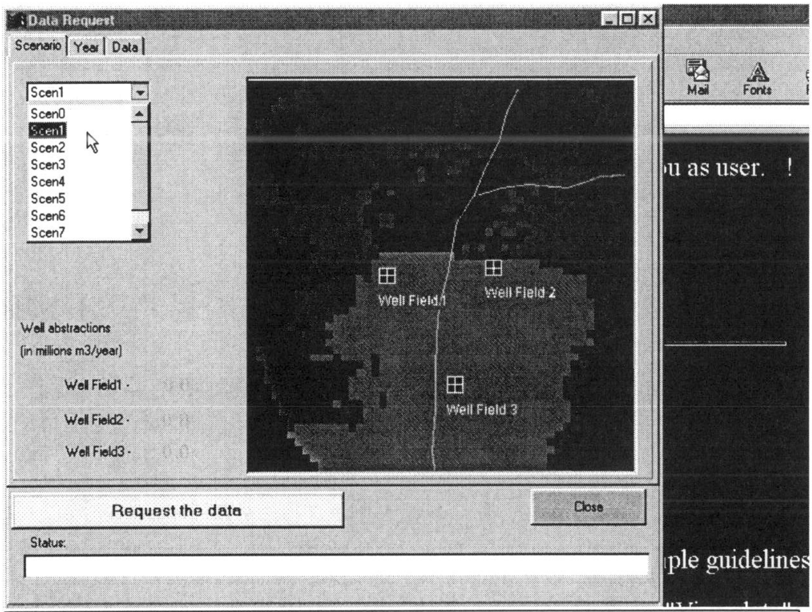

Figure 4.16. Request Data - Step 1: Choosing scenario

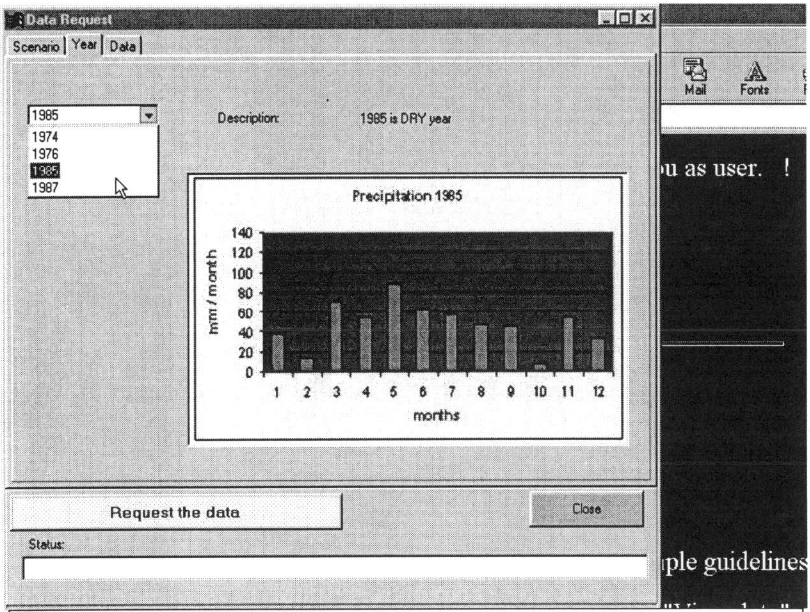

Figure 4.17. Request Data - Step 2: Choosing hydrological year

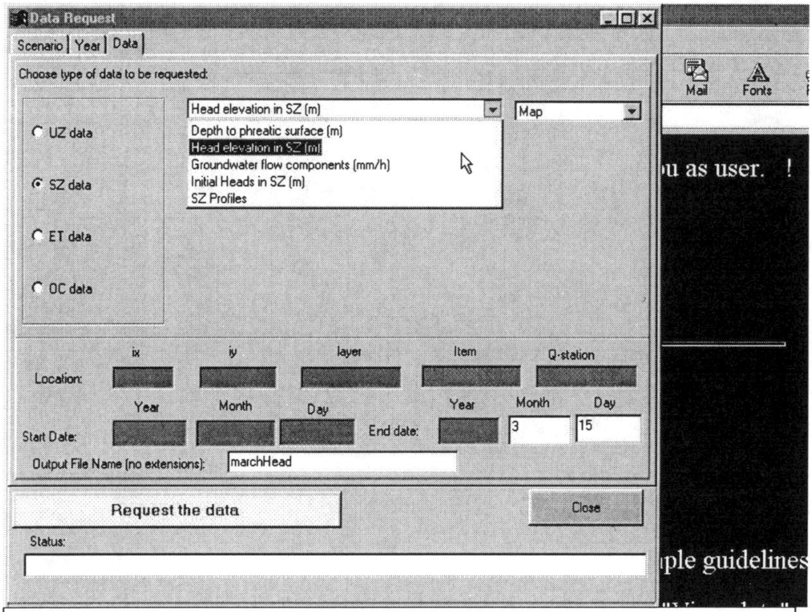

Figure 4.18. Request Data - Step 3: Choosing type of data to be requested

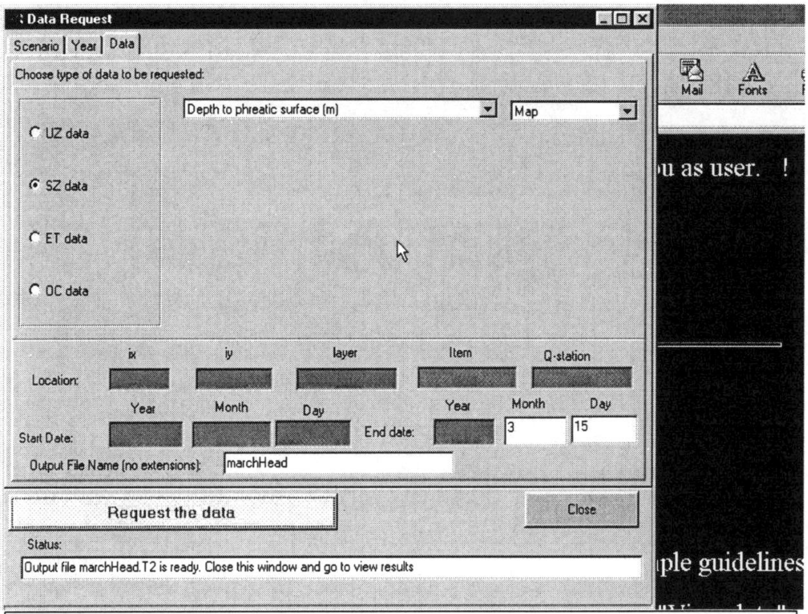

Figure 4.19. Message from server reporting the conclusion of the request

After choosing the scenario and the hydrological year for which the data are going to be requested, the user is led to an interface for requesting the actual data. Because nearly all *input* data are common to all users they are accessible through the users directory created on the server without formulating special requests. Only some specific model inputs need to be requested by the users, while most of the data that are actually requested are model outputs. The formulated request is sent to one of the servlets in the servlet layer, depending on whether it is input or output data (IR – Input retrieval; OR - Output Retrieval). The servlet programs then open a low-level socket connection with a server program that wraps the MIKE SHE executable IR and OR programs. MIKE SHE uses these programs for retrieval of model data from two binary files storing the model inputs (the 'flow input file' - *fif*) and the model results (the 'flow result file' – *frf*). The programs extract the requested data and store them in the user's directory as an ASCII file. When the procedure is concluded a message is sent to the client that the request is finished and that the data can be seen through the second frame: 'View Data'.

Table 4.2 shows all the input and output data that are available and the ways in which they can be presented.

Table 4.2 Available model data

Type of data	**Map**	**Time Series**	**Special format-comment**
Model Input			
General			
Precipitation		*	
Topography	*		
Vegetation Distribution	*		
Soil Distribution	*		
Paved areas	*		
Catchment Boundary	*		
Overland Flow			
Detention Storage	*		
Manning coefficient	*		
Unsaturated zone			
UZ classification codes	*		
UZ profiles			Special graphics
Soil characteristics			Special graphics
Saturated zone			
Boundary grid codes	*		
Horizontal hydraulic conductivity	*		
Vertical hydraulic conductivity	*		
Specific yield	*		
Storage coefficient	*		
Boundary flow gradient in x-direction	*		
Boundary flow gradient in y-direction	*		
Initial groundwater heads	*		Depending on scenario
SZ profiles			Special graphics

Table 4.2 – continued

Type of data	Map	Time Series	Special format-comment
Model output			
Evapotranspiration data	*	*	
Actual evapotranspiration	*	*	
Actual transpiration	*	*	
Evaporation from soil surface	*	*	
Evaporation from intercepted storage	*	*	
Evaporation from ponded water	*	*	
Canopy storage	*	*	
Infiltration to Unsaturated zone	*	*	
Evapotranspiration from saturated zone	*	*	
Overland and channel flow data			
Depth of overland flow	*	*	
Overland flow in x an y directions	*	*	
Water depth in river at Q-stations		*	
Flow in river at Q-stations		*	
Total inflow in river from overland		*	
Total inflow in river from aquifers		*	
Total inflow in river from drains		*	
Unsaturated zone data			
Rate of change of storage in UZ	*	*	
Recharge to Saturated Zone	*	*	
Flow in UZ			Special format
Soil moisture content in UZ			Special format
Saturated zone data			
Depth to phreatic surface	*	*	
Elevation of groundwater head	*	*	
Groundwater flow in x and y directions	*	*	

The second frame ('View Data') is used for the actual data presentation. Upon loading of this frame, the updated status of the user's directory (including all recently requested files) are sent to the client-side from where the user can access all common and requested data. The *UpdateServlet* is responsible for this task. By selection of a particular data file, the *SendDataServlet* reads the relevant ASCII file and sends the data to the client-side. After receiving the data, the frame produces plots in the appropriate format (as maps or time series). The interface for plotting maps is built with support for zooming, display of the model grid, the well fields, and the rivers with the Q stations (river nodes). The range of values and the number of classes in which these values are presented in the map are adjustable. Four different colour schemes are implemented. All the maps from the MIKE SHE model are easily displayed with this interface. The time series interface plots the data with an automatic scaling of axes that depends upon the data received. The frame also supports the display of other kinds of data. The interface supports the special MIKE SHE-kind of display of UZ data (soil moisture content or UZ flow, see Table 4.2) in graphs where the vertical axis pointing downwards represents the depth, and the horizontal axis represents time, while the actual data are presented as colour distributions. It is also possible to provide graphical presentations of UZ profiles including all soil characteristics, SZ profiles, and river node data. Figures 4.20-4.25 show some examples of the user interface provided for the viewing of data.

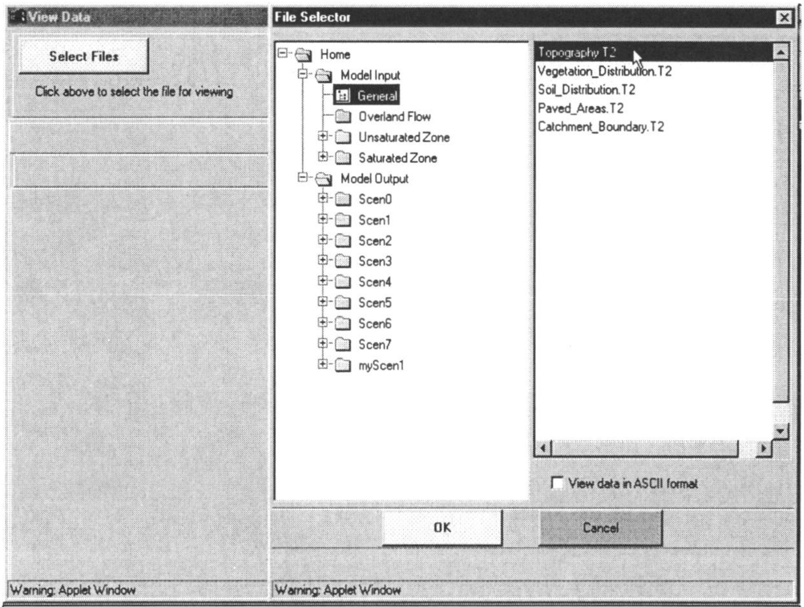

Figure 4.20. View Data – Interface to the user directory

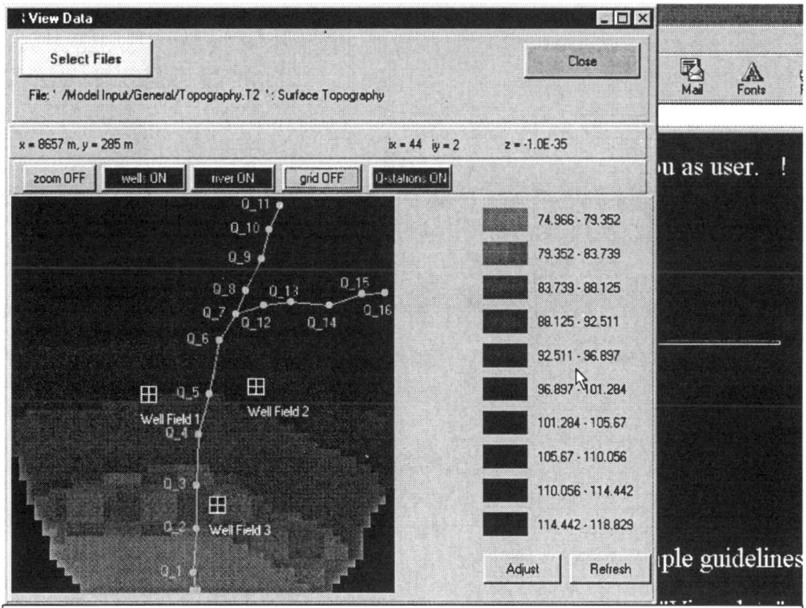

Figure 4.21. View Data – Map interface

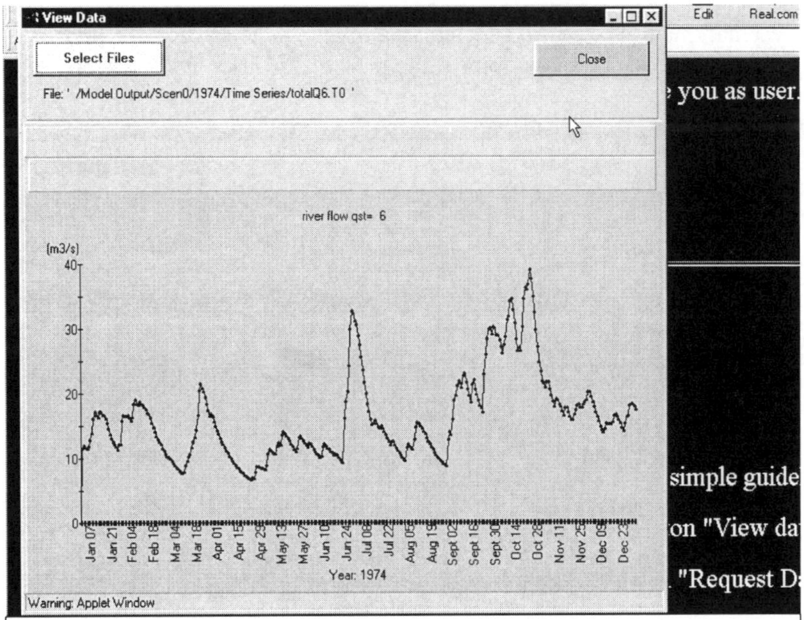

Figure 4.22. View Data - Time series interface

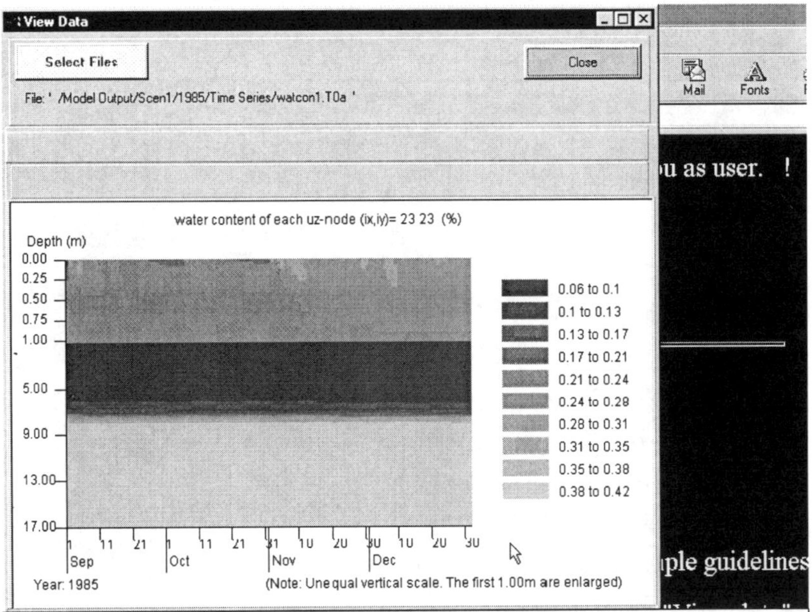

Figure 4.23. View Data – Interface for viewing UZ data over depth/time

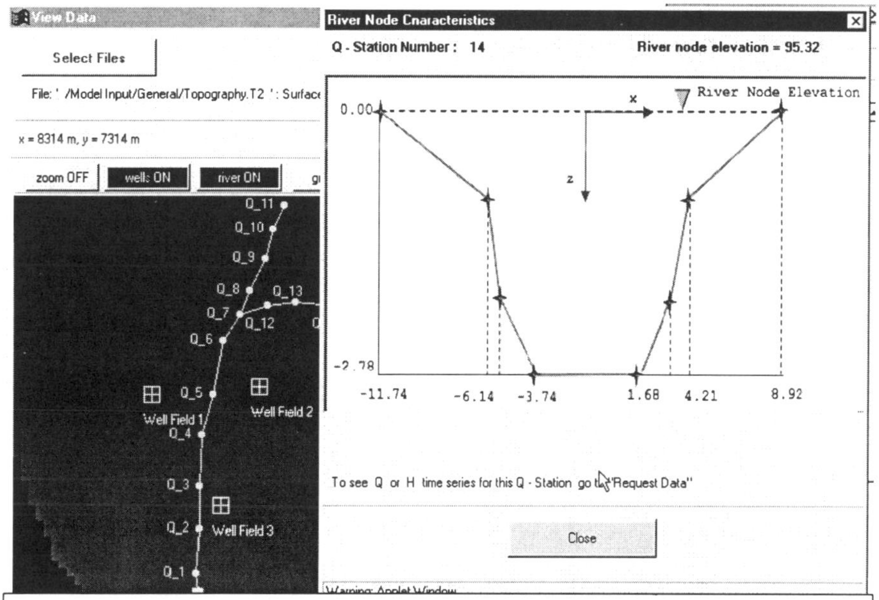

Figure 4.24. View Data – Interface for river nodes data

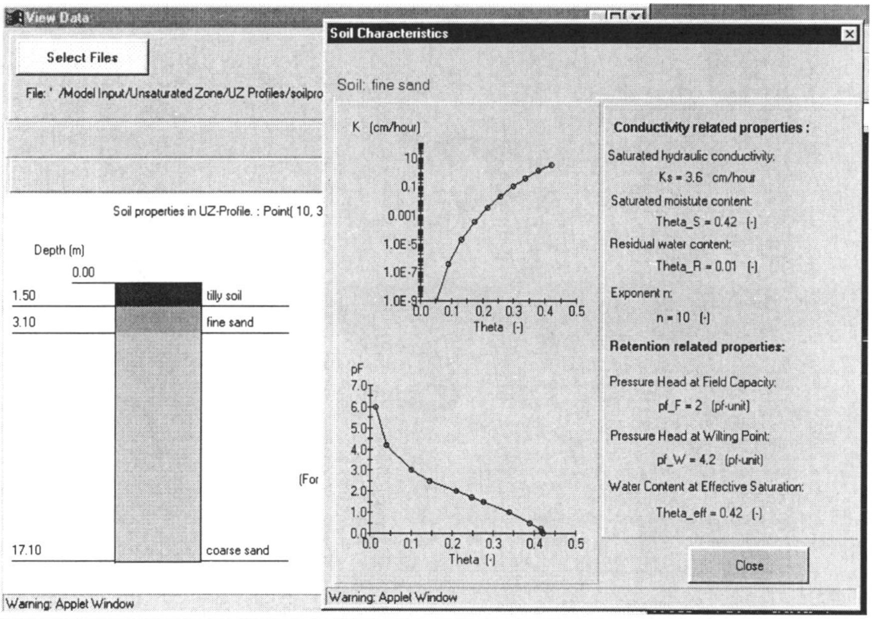

Figure 4.25. View Data – Interface for soil profiles and soil characteristics

The third frame of the user interface – 'New Scenario' - allows for running user-specified scenarios, other than the 8 already made available. This feature is not yet included specifically in the *Aquavoice* game, but may be included in its future development. It may serve, however, as an example of the possibilities for extending such applications by actually running a MIKE SHE simulations remotely, over the Internet. Through this frame (Figure 4.26 and 4.27) the user may request a new MIKE SHE simulation for different combinations of abstractions from the three potential well fields. With the new data, a MIKE SHE configuration is prepared (from new wells data, new data about initial heads, etc.), and then the actual simulation is started. The progress of the simulation is reported to the client. Two servlets are responsible for handling these two tasks: the *SetupWMServlet* and the *RunWMServlet*. They are connected respectively with two servers that wrap the MIKE SHE executables again through the low-level socket connections. The interface provides means for stopping a running simulation at any moment, with the results being available up until the stopping moment. After finishing, the results from the new scenario are stored on the server side, and accessible from the 'Request Data' and 'View Data' frames in the same way as for the existing 8 scenarios. The *UpdateServlet* ensures that the most recent status for the particular user is always presented in the interface.

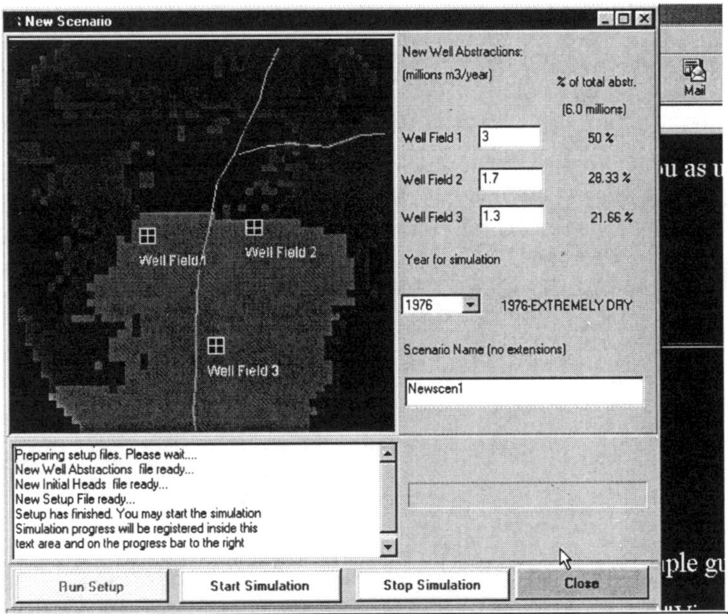

Figure 4.26. New Scenario – Setup for a new simulation

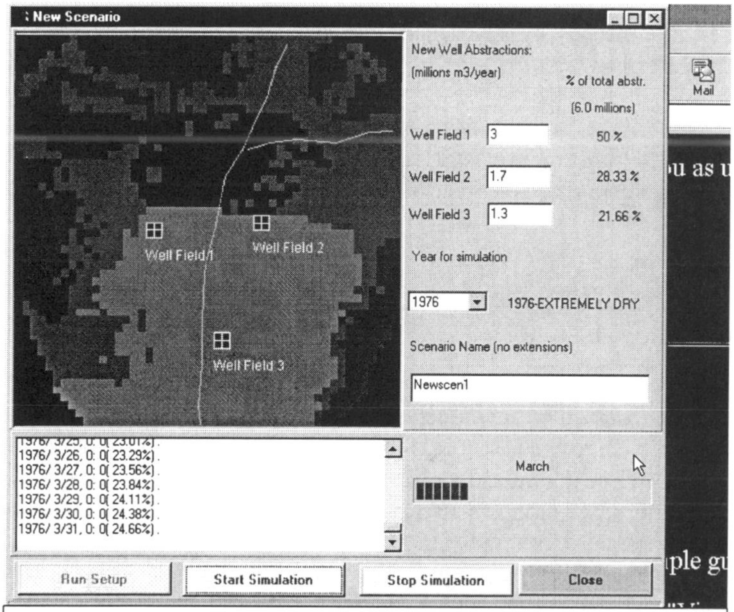

Figure 4.27. New Scenario – Running new simulation

The MIKE SHE remote simulation is designed in such a way that several different users can run it concurrently. This is achieved by using multithreading. The *RunWMServlet*, serves a collection of the clients with simulation requests and keeps a separate thread of execution in its communication, with the multithreaded server running the main MIKE SHE executable program[13]. The limits of the number of users running MIKE SHE simulations concurrently will obviously be determined by the capacity of the machine(s) where the MIKE SHE servers run. The solution to link all servers that wrap the MIKE SHE executables with low level socket connections was realised exactly for the purpose of separating all MIKE SHE related tasks from the Web server, which can be overloaded with all kinds of other requests. This software solution can be implemented advantageously when using separate machines dedicated to MIKE SHE.

This last feature of this application shows how problems of remote modelling can be approached *in principle* through such a framework. Because of the problem at hand, however, only the well abstractions are made adjustable in *this* application. In other kinds of problems, some different kinds of inputs may be provided to the users for manipulation. The same implementation architecture allows for extending even the whole modelling procedure over the WWW if that is needed. Within the context of NDDSSs, however, the design of fact engines will generally be limited to instantiating certain models and then exposing only a limited number of inputs (namely those related to the decision problem at hand). These will

[13] The multithreading approach is used only for the remote MIKE SHE simulation because this is the most time demanding task. For all other tasks in this application, the servlets and the back-end servers run as single threads.

be manipulated either directly by the users of the system or through judgement engines of the kind that will be described next.

The *Aquavoice* game, in its present form, deals with a choice decision problem, and only the results of the 8 pre-existing scenarios are used. The following section will describe how these results are used for supporting the judgement formation process of the participants in an *Aquavoice* session.

4.2.2 The Judgement Engine

In order to present the design of the judgement engine component of *Aquavoice* we now need to turn our attention back to the actual roles of the players participating in the game. Each of the five basic roles (Local Authority, Environmental Agency, Farmers' Council, Water Supply Company and any Individual user) uses the judgement engine component to judge and evaluate the proposed scenarios of groundwater abstraction. *Aquavoice* uses only the 8 pre-run scenarios, and then only for an average hydrological year (1974). The goal of the judgement engine is to allow each user, in whatever role he or she is participating, initially to create an *individual judgement and evaluation* of the proposed scenarios. The aggregation of these individual evaluations is then utilised by the participants through the intervention of the negotiation and collaboration platform.

The judgement engine of *Aquavoice* is built by taking the approach that each role first uses the interface of the system to view some data resulting from the fact engine (the MIKE SHE model). Then, on the basis of these facts, each role-player individually *judges* the proposed scenarios. As mentioned earlier, the use of pre-defined procedures for impact assessment has been avoided, allowing the users to create their judgements independently. It is envisaged that such procedures, which contain expert knowledge about the actual impacts, can be encapsulated into separated components, and provided to the users within the judgement engine environment, but separated from the actual application (for example, in the help pages). In this way the expert advice about the impacts can be provided, but the users remains free to choose whether to follow that advice or not.

All the roles use linguistic terms for forming their judgements of the proposed scenarios with respect to the issue of their control, except the Water Supply Company, which uses a separate economic model for the same purpose. At, the same time, however, each user has to *evaluate* the scenarios with respect to four main issues. Consequently the procedure for creating judgements and evaluations is designed on the basis of one of the currently popular Fuzzy Multiple Attribute Decision Making Methods (FMADMs). Before looking at the actual adaptation and implementation of this procedure, and already in the first subsection, we will introduce the method itself. The second subsection will describe the adaptation of the method for the purposes of *Aquavoice*. As the actual implementation is inseparable from the overall design of *Aquavoice* and underlines the dynamics of the game, it will be presented in the next section (4.3).

4.2.2.1 Description of the FMADM adapted for the *Aquavoice* judgement engine

Depending on the type of decision problem at hand, the numerous methods of the broad field of Multi Criteria Analysis (MCA) can broadly be classified in two large groups: the so-called Multiple Objective Decision Making Methods (MODM) and Multiple Attributes Decision Making Methods (MADM) (Hwang and Masud, 1979; Hwang and Yoon, 1981). The first group of methods addresses problems with multiple criteria (here called 'objectives') where the number of potential alternatives is not predetermined, and most generally can be assumed as infinite. The methods are used to *design* (or search for) one alternative that will be considered as 'best' or 'optimal'. All classes of multi-objective optimisation methods belong to this group. The second group, on the other hand, addresses problems in which the number of alternatives in the decision problem is predetermined and finite. These methods are used for *choosing* (evaluating) the predetermined alternatives on the basis of multiple criteria (here called 'attributes').

The method implemented in *Aquavoice* belongs to the second group, where the *alternatives* considered are the 8 proposed *scenarios* of groundwater abstraction, and the *attributes* are the *issues* raised by the participants in the game. The introduction of these alternative terms in this study may seem unnecessary for readers familiar with standard MADM terminology, but it has been found more suitable within the *Aquavoice* context, and especially for its potential users (most of them probably lacking expert MCA knowledge). While in this subsection we will use the terms *alternatives* and *attributes* for introducing the method, in the next subsection, which introduces the adaptation of the method to *Aquavoice*, we will switch to the alternative terminology.

We will follow the common form of expressing a MADM problem using the so-called decision matrix (D). D is an (*mxn*) matrix whose element x_{ij} indicate the evaluation or 'attribute value' of alternative i (A_i), with respect to attribute j (X_j). In this notation it is assumed that there are *m* alternatives (A_i, i=1,2....m) and *n* attributes (X_j, j=1,2...n). Thus the matrix has the form as in Table 4.3:

Table 4.3 General form of a decision matrix

	Attributes					
Alternatives	X_1	X_2	...	X_j	...	X_n
A_1	x_{11}	x_{12}	...	x_{1j}	...	x_{1n}
A_2	x_{21}	x_{22}	...	x_{2j}	...	x_{2n}
...	...	...	...	...	...	...
A_i	x_{i1}	x_{i2}	...	x_{ij}	...	x_{in}
...	...	...	...	...	...	...
A_m	x_{m1}	x_{m2}	...	x_{mj}	...	x_{mn}

There are many criteria that can be used for classification of the various MADM methods. A detailed overview is given in Hwang and Yoon, 1981 and Chen and Hwang, 1992. Without going into details we will here introduce some of the main general aspects of MADM problems which are relevant to the method used here.

The attributes of an MADM problem may be *cost* or *benefit* attributes. Larger attribute values are preferred for benefit attributes, and smaller attribute values are preferred for cost attributes. By using the notion of *utility function* (which can be compared to the value functions introduced earlier), considering the given attribute as cost or benefit means that the attribute has a *monotonic* utility function. Cost attributes have monotonically decreasing utility, and benefit attributes have monotonically increasing utility. Attributes can have non-monotonic utility functions as well (such as square shape or bell-shape utility functions) in which case they cannot be considered as costs or benefits. In most MADM methods the utility function is only implicitly represented.

An important notion in MADM problems is the one of an *ideal solution* (sometimes called an optimal solution or 'utopia point'). This is a solution of an imaginary alternative which has maximum values for all benefit attributes, and minimum values for all cost attributes. Or, using the notion of utility this solution maximises the utility of all attributes simultaneously. This solution does not actually exist (or if it exists there are no conflicting attributes, and the problem of choice disappears), but it is very important since many methods obtain a solution from the existing alternatives based on its *closeness* to this ideal solution.

Many MADM methods also require additional information about the relative importance of each attribute. In the so-called compensatory methods, which allow for tradeoffs between attributes and are among the most common methods, this information is usually given by a set of preference *weights* which are normalised to sum to 1. Several methods have been developed for assessing these weights on the basis of the information already contained in the decision matrix (i.e. by trying to 'objectivise' the procedure of assigning weights), but very often the weights are directly given by the 'decision maker' and reflect personal preferences.

Traditional MADM methods assume that all attribute values are quantitative, and indeed expressible by crisp numbers. In such cases, the only problem that needs to be resolved by the method is the presence of the unequal units of measurement used for the different attributes. This is resolved by the so-called *normalisation* of the attribute values (involving the reduction of all attributes to one comparable scale, usually between 0 and 1).

In many cases, however, attribute values are supplied as qualitative information only. The introduction of fuzzy set theory into MADM methods allows the conversion of such 'vague' information into quantitative information without resorting to some quite arbitrary methods (such as using some interval scales) that have been used earlier. With fuzzy sets it is possible to deal with qualitative, vague, attribute values directly. In general the decision matrix will then contain a mixture of fuzzy and crisp attribute values. This is the case of the *Aquavoice* choice problem as well, and the method adapted here assumes the relevance of such a mixed decision matrix.

The method employed here was proposed by Chen and Hwang, (1992), as being more efficient then other existing FMADM methods. Assuming that the decision matrix contains some attribute values expressed in linguistic terms and other in crisp numbers, this method first converts all the linguistic terms into fuzzy sets (of numbers), and then from these fuzzy sets calculates the so-called crisp *scores* for each linguistic term used. After this procedure, the decision matrix contains only crisp numbers and the problem is solved by any of the existing MADM methods. For this purpose the TOPSIS method (Technique for Order Preference by Similarity to Ideal Solution) is recommended. Let us first present the procedure for converting the linguistic terms into crisp scores using fuzzy sets.

The total universe used consists of 13 different linguistic terms: *none, very low, low to very low, low, medium low, more or less low, medium, more or less high, medium high, high, high to very high, very high and extremely high.* The systematic conversion of these linguistic terms into fuzzy numbers is achieved by making use of 8 different conversion scales. Each scale contains different combination of linguistic terms, varying from a simple combination of only two terms (in scale 1) up to refined gradual variation expressed with eleven terms (in scale 8). The terms are mapped into fuzzy sets with different membership functions as presented in Figure 4.28a to 4.28h. The membership functions of all fuzzy sets used $\mu_M(x)$ are in the closed interval [0,1]. The support of all the fuzzy sets used (values of x with non-zero membership) belongs to the universe of the set of real numbers in the interval [0,1]. These fuzzy sets are defined as fuzzy numbers.

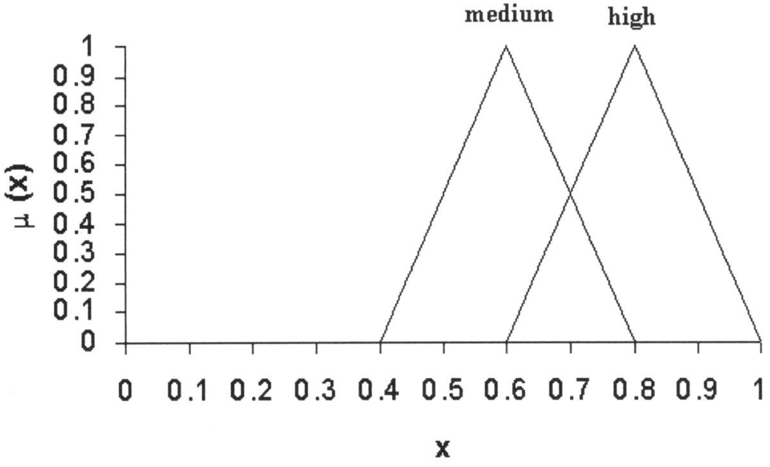

Figure 4.28a. Linguistic to fuzzy conversion: Scale 1

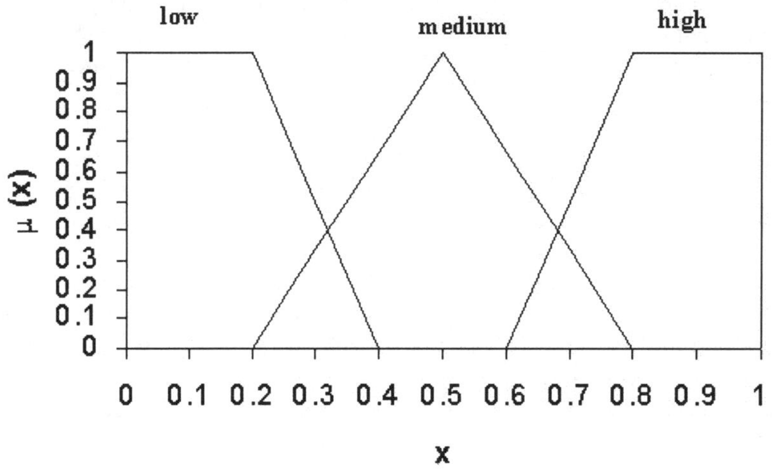

Figure 4.28b. Linguistic to fuzzy conversion: Scale 2

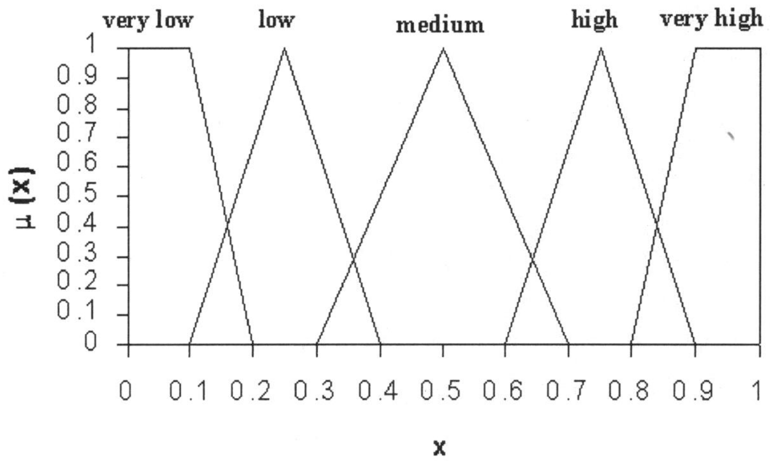

Figure 4.28c. Linguistic to fuzzy conversion: Scale 3

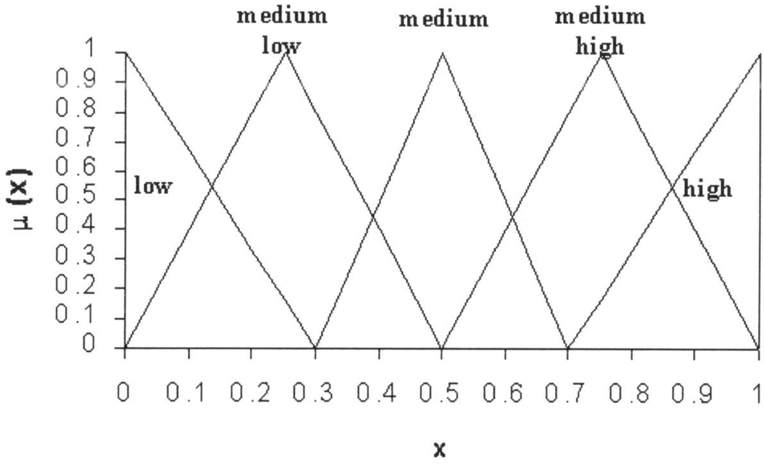

Figure 4.28d. Linguistic to fuzzy conversion: Scale 4

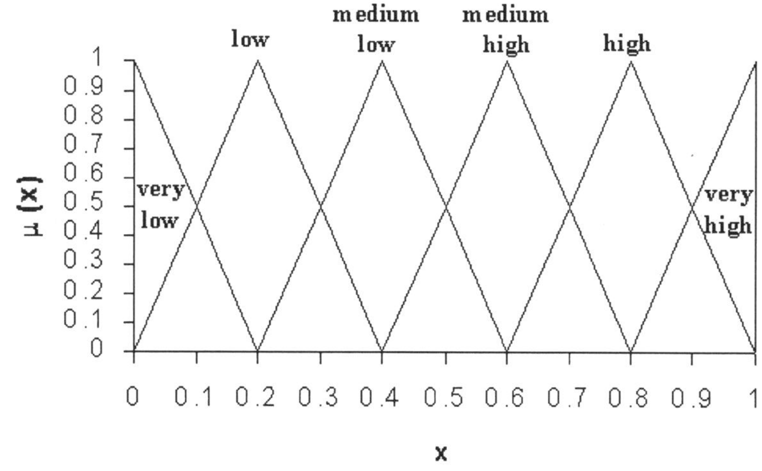

Figure 4.28e. Linguistic to fuzzy conversion: Scale 5

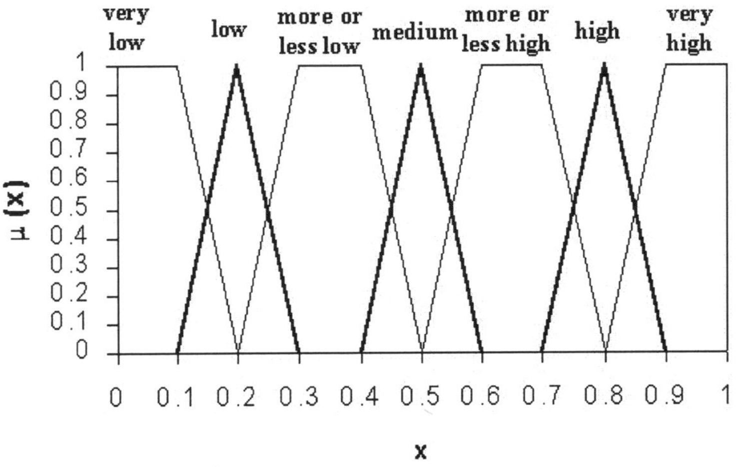

Figure 4.28f. Linguistic to fuzzy conversion: Scale 6

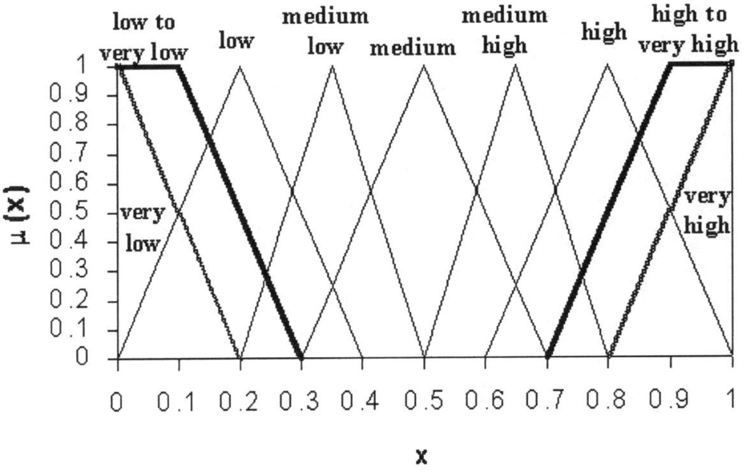

Figure 4.28g. Linguistic to fuzzy conversion: Scale 7

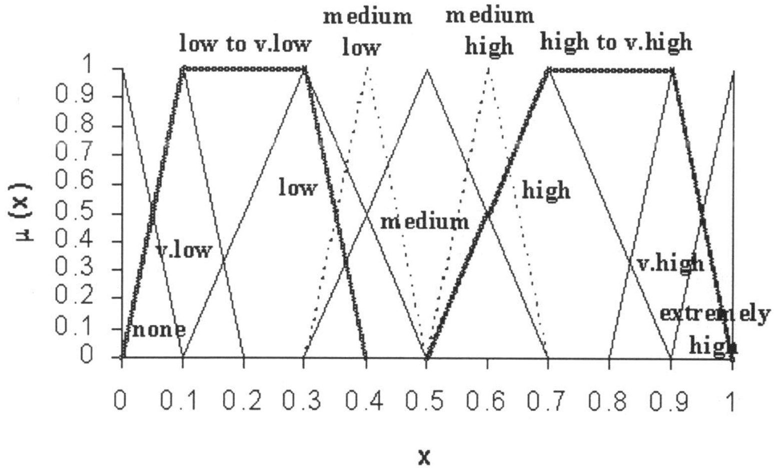

Figure 4.28h. Linguistic to fuzzy conversion: Scale 8

The linguistic terms used here need to be understand as generic, in the sense that, for different attributes, different pair of words with extreme meanings can be found with appropriate terms expressing the intermediate grades (good vs. poor; low vs. high; small vs. large etc.)

The principle of the system is to pick up a scale that contains all the terms used for expressing the values of a given attribute, while always using the scale with the smallest number of terms. For example, if the only two terms used are *medium* and *high*, scale 1 will be chosen, regardless of all other scales containing such terms. On the other hand, if the terms used are *medium*, *high* and *extremely high*, scale 8 will be chosen because it is the only one containing *extremely high.* In this way the linguistic terms are converted into fuzzy numbers.

The crisp 'scores' of these fuzzy numbers will depend on the actual scale which is chosen because same terms have different membership functions in different scales. The actual conversion of the fuzzy numbers into crisp scores uses the fuzzy minimising set and the fuzzy maximising set as defined in Figure 4.29. The maximum membership value of the intersection of the minimising set and the membership function of the fuzzy number M is the so-called *left score* of M. The maximum membership value of the intersection of the maximising set with the membership function of the fuzzy number M is the *right score* of M. The total crisp score of a given fuzzy number is average of these two scores calculated as indicated in Figure 4.29.

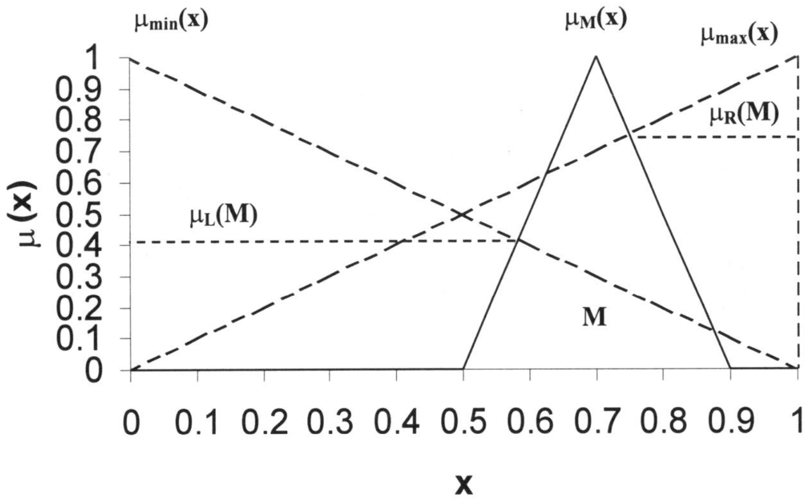

Fuzzy number M with membership function $\mu_M(x)$

Maximising set:

$$\mu_{max}(x) = \begin{cases} x, & 0 \le x \le 1 \\ 0, & \text{otherwise} \end{cases}$$

Minimising set:

$$\mu_{min}(x) = \begin{cases} 1-x, & 0 \le x \le 1 \\ 0, & \text{otherwise} \end{cases}$$

Left score:

$$\mu_L(M) = \max \mu(x) \text{ of } [\mu_M(x) \cap \mu_{min}(x)]$$

Right score:

$$\mu_R(M) = \max \mu(x) \text{ of } [\mu_M(x) \cap \mu_{max}(x)]$$

Total score:

$$\mu_T(M) = [\mu_R(M) + 1 - \mu_L(M)] / 2$$

Figure 4.29. Calculation of the crisp score of a given fuzzy number

With this procedure, the crisp score of each fuzzy number from any of the eight conversion scales is calculated. Once the original linguistic terms are converted into crisp numbers, the TOPSIS method is used for the actual ranking of the alternatives. Because in general the decision matrix contains mixed kinds of attribute values, after conversion of the linguistic terms into crisp scores it is usually necessary to normalise the decision matrix in order to obtain comparable scales. TOPSIS uses the so–called vector normalisation. Each attribute value is divided by the Euclidean norm of the vector of the attribute at hand. The normalised decision matrix is then converted into a weighted normalised matrix using the set of attribute preference (or importance) weights, normalised to sum to 1, which are provided by the 'decision maker'.

The essence of the method is that it uses cost and benefit attributes, which means that it can only work with monotonic utility functions. The 'decision maker' determines which are cost and which are benefit attributes according to the problem at hand. In such a case it is easy to locate the 'positive-ideal solution' (all the 'best' attribute values – maximum for benefits and minimum for costs) and the 'negative-ideal solution (all the 'worst' attribute values – minimum benefits and maximum costs). The existing alternatives are then ranked according to their *relative* closeness to the ideal solution. The steps of the TOPSIS method, including the calculation of the relative closeness are summarised in Figure 4.30.

Decision matrix D (*mxn*) with crisp attribute values

m - number of alternatives
n - number of attributes

***Step 1*: Calculate the normalised decision matrix R. Normalised attribute value r_{ij} is calculated as:**

$$r_{ij} = \frac{x_{ij}}{\sqrt{\sum_{i=1}^{m} x_{ij}^2}}, \quad i = 1,2,\ldots,m;\ j = 1,2,\ldots,n.$$

***Step 2*: Calculate the weighted decision matrix V. The weighted normalised attribute value v_{ij} is calculated as:**

$$v_{ij} = w_j r_{ij}$$

where w_j is the weight of the j-th attribute and

$$\sum_{j=1}^{n} w_j = 1$$

Step 3*: Determine the attributes as benefits *J* or costs *J'

***Step 4*: Calculate the alternatives called positive ideal solution A^+ and negative ideal solution A^- as:**

$$A^+ = \{(\max_i v_{ij} \mid j \in J), (\min_i v_{ij} \mid j \in J') \mid i = 1,2,\ldots,m\} = \{v_1^+, v_2^+, \ldots, v_j^+ \ldots v_n^+\}$$

$$A^- = \{(\min_i v_{ij} \mid j \in J), (\max_i v_{ij} \mid j \in J') \mid i = 1,2,\ldots,m\} = \{v_1^-, v_2^-, \ldots, v_j^- \ldots v_n^-\}$$

***Step 5*: Calculate the separation measures between each alternative and the two respective ideal solutions by Euclidian distance:**

Separation from A⁺: $$S_i^+ = \sqrt{\sum_{j=1}^{n} (v_{ij} - v_j^+)^2}$$

Separation from A⁻: $$S_i^- = \sqrt{\sum_{j=1}^{n} (v_{ij} - v_j^-)^2}$$

***Step 6*: Calculate the relative closeness to ideal solution for each alternative:**

$$C_i = \frac{S_i^-}{S_i^+ - S_i^-}$$

Figure 4.30. Steps of the TOPSIS method

The measure of relative closeness has values between 0 and 1. Alternatives with values approaching 1 are closer to the positive-ideal solution, and those with values approaching 0 are closer to the negative-ideal solution. As a result of this procedure, the alternatives are ranked in descending order according to the values of their relative closeness to the ideal solution.

4.2.2.2 Adaptation of the FMADM for the *Aquavoice* judgement engine

The method described above was originally designed for determining of the ranking of the proposed alternatives by *one* 'decision maker'. In principle it was assumed that the provision of the attribute values, the assignment of the preference weights and the classification of the attributes involved as costs or benefits were all the responsibilities of that 'decision maker'.

In *Aquavoice*, each participant in the game uses the same method for obtaining the individual ranking of the alternatives. However, the adaptation of the method for the *Aquavoice* judgement engine takes a slightly different approach again. As we have already mentioned, in *Aquavoice* the *alternatives* are the 8 *scenarios* of groundwater abstraction, and the *attributes* are the *issues* controlled by the four main institutional roles. The attribute values are consequently called *issue judgements.* Thus, for these four institutional roles the decision matrix will consist of eight scenarios and four issues. For the role of individual user, if he or she decides to raise an additional issue, the number of issues may be five.

The whole procedure in *Aquavoice* is divided into two stages. The first stage is called the *judgement* stage, and the second is called the *evaluation* stage. If we compare this with the method described in the previous subsection we can say that the judgement stage corresponds to the provision of attribute values (issue judgements), and the actual formulation of the decision matrix. This stage, as we shall see, is carried out concurrently by the four institutional role players in such a way that the basic formulation of the 8x4 decision matrix is common to all players/participants. The second, evaluation, stage is carried out by each player individually. It consists of classifying the issues as costs or benefits, and by assigning the preference weights. Because these two steps of the evaluation stage are carried out independently by each user, the individual rankings of the 8 scenarios will in general be different for each user.

We will first describe the judgement stage. Since the four main issues are 'controlled' by the four respective institutions (see again page 147), the provision of the issue judgements for a given issue is allowed only for the role player who is responsible for that issue. For example, only the Local Authority can provide issue judgements for the issue 'Citadel' and only the Environmental Agency can provide issue judgements for the issue 'Forest'. However, the issue judgements provided by one role player enter as columns of the common decision matrix of every player/participant.

Most of the judgements are provided by the role players as linguistic terms. The universe of terms used consists of 11 terms: *extremely bad, very bad, bad to very bad, bad, fairly bad, medium, fairly good, good, good to very good, very good* and *excellent*. The conversion of these 11 terms into crisp scores is done with the 8 conversion scales. These terms coincide with the 13 terms proposed by the FMADM method with the difference that the terms *fairly bad* and *fairly good* coincide both with the pair *medium low* and *medium high* (used in scales 4,5,7 and 8), and with the pair *more or less low* and *more or less high* (used in scale 6).

The best way to describe the procedure in the judgement stage is by following the steps taken by each of the four institutional roles:

Local Authority

Through the user interface of the system, the Local Authority (LA) receives a restricted set of facts provided as outputs of the MIKE SHE model. Because the issue of concern for the LA is the influence of groundwater abstractions on the Citadel, the role player can choose to see some or all of the following data:

-Time series of depth to phreatic surface in the middle of the Citadel area
-Maps of depth to phreatic surface in the dry period (and specifically on September 15, as the most critical day in the dry period)

By looking at the data provided for all 8 scenarios, the role player provides issue judgements with respect to the issue of 'Citadel' for each scenario. All issue judgements are provided in linguistic terms. This is the first column in the common decision matrix, which is valid for all players/participants.

Environmental Agency

Being responsible for the issue 'Forest', the Environmental Agency looks at the following facts provided from the MIKE SHE model:

-water content in the unsaturated zone during the summer in a grid cell in the forest
-maps of depth to phreatic surface in the dry period (and specifically on September 15)
-time series of depth to phreatic surface in a grid cell in the forest

Following the same procedure, it provides the issue judgements as linguistic terms. This is the second column of the decision matrix.

Farmers' Council

The Farmers Council is responsible for the issue 'Cornfields'. It looks at the following data:

-water content in the unsaturated zone during the summer in a grid cell in the cornfields
-maps of depth to phreatic surface in the dry period (on September 15 again)
-time series of depth to phreatic surface in a grid cell in the cornfields

Again, the issue judgements are provided in linguistic terms, and the third column in the decision matrix is formulated.

Water Supply Company

The water supply company is responsible for controlling the issue of 'Water Price'. It does not use the MIKE SHE model data, but has at its disposal a simple economic model based on a cost-benefit analysis that calculates the price for certain target profit levels. The model is set-up in such a way that each potential well field brings its own installation and operation costs, which are however also dependent on the quantity of abstracted water. As the cost-benefit analysis depends crucially on the duration time of the project, one very important parameter in the model is the duration of the abstraction licence provided by the Environmental Agency. The role player actually has an interface for manipulating different water prices and obtains different profits correspondingly. Because of the different costs and the different licence durations, each scenario will eventually result in a different water price and different profit. The combination water price / profit will eventually depend on the choice of the role player.

The model actually calculates the 'profit' as being equal to the so-called Net Present Value (NPV), a concept often used in capital investment analysis for evaluating new projects. The NPV is the *present* value of all cash flows in a given project, both costs and benefits. Here, for each scenario the NPV is calculated as the difference between the present value of the benefits and the present value of the costs. All the terms used in the model are explained in Figure 4.31.

As shown in Figure 4.31 apart from all the costs directly related to the scenarios of groundwater abstraction, the Water Supply company may offer some compensations to the Farmers Council during the course of the game in order to persuade them to accept a certain scenario. This term also enters into the model as a cost term. All parameters for the model are assumed in such a way that 'reasonable' results for water price and for profits are obtained As a result of these calculations, each scenario sets a value for the water price. These values provide the fourth column in the common decision matrix.

Net present value: $NPV = PV_B - PV_C$

PV_B - **Present value of benefits**
PV_C - **present value of costs**

Benefits

$$PV_B = \sum_{k=1}^{n} \frac{Q_{tot} P_w K_{co}}{(1+d)^k}$$

Q_{tot} - **Total abstraction rate (6x10^6 m^3/year)**

P_w - **price of water per 1 m3**

K_{co} - **coefficient describing the amount (in %) actually billed *and* collected (assumed 70%)**

n - abstraction licence duration (in years)

d - discount rate (assumed 7%)

Costs

$$PV_C = C_0 + C_1 + C_2$$

C_0 - **fixed installation costs depending on scenario of abstraction**

C_1 - **fixed costs for compensations offered to the Farmers Council**

C_2 - **Operational costs depending on time and scenario of abstraction calculated as:**

$$C_2 = \sum_{k=1}^{n} \frac{Q_{tot} C_{op}}{(1+d)^k}$$

C_{op} - **coefficient of operational costs depending on scenario of abstraction**

Figure 4.31. Calculation of Net Present value for each scenario

When the four institutional players/participants have provided their judgements, the basic common decision matrix has the form exemplified in Table 4.4:

Table 4.4 Example of formulated common decision matrix

	Issues			
	Citadel	*Forest*	*Corn fields*	*Water price*
Scenario	Provided by LA (linguistic)	Provided by EA (linguistic)	provided by FC (linguistic)	provided by WS (crisp; euro/ m^3)
Scenario 0	Fairly bad	Good	Good	0.5
Scenario 1	Fairly good	Extremely bad	Excellent	0.55
Scenario 2	Very bad	Good	Medium	0.60
Scenario 3	Good	Excellent	Extremely bad	0.45
Scenario 4	Bad	Bad	Very good	0.70
Scenario 5	Fairly bad	Fairly good	Very bad	0.55
Scenario 6	Very good	Good	Very bad	0.65
Scenario 7	Very good	Fairly bad	Bad	0.75

Every player/participants will have to evaluate the proposed scenarios *at least* against these four issues using the above matrix. Of all the institutional players, the only role that has a slightly different matrix is the Water Supply Company. This is because, from the economic model, results are obtained both in terms of price of water and obtained profit, so that this role will have one more column in its decision matrix, representing the issue of 'profit' (see Table 4.5).

Table 4.5 Example of extended decision matrix for Water Supply Company

	Issues				
	Citadel	*Forest*	*Corn fields*	*Water price*	*Profit*
Scenario	Provided by LA (linguistic)	Provided by EA (linguistic)	provided by FC (linguistic)	provided by WS (crisp; euro/ m^3)	Provided by WS (crisp; 10^6 euro)
Scenario 0	Fairly bad	Good	Good	0.5	0.0
Scenario 1	Fairly good	Extremely bad	Excellent	0.55	4.7
Scenario 2	Very bad	Good	Medium	0.60	5.5
Scenario 3	Good	Excellent	Extremely bad	0.45	5.3
Scenario 4	Bad	Bad	Very good	0.70	7.9
Scenario 5	Fairly bad	Fairly good	Very bad	0.55	4.3
Scenario 6	Very good	Good	Very bad	0.65	5.6
Scenario 7	Very good	Fairly bad	Bad	0.75	3.1

Individual user

Players/participants who take the role of individual users in *Aquavoice* first need to choose whether they wish to evaluate the scenarios using the common decision matrix created by the four institutional players, or whether instead they want to raise an additional issue of their own. In the first case they can use the common decision matrix and move directly to the evaluation stage, which will be discussed below. In case they choose to raise an additional issue, they will first have to formulate that issue. Several issues are pre-provided in *Aquavoice*, such as those appertaining to different agricultural crops, but the individual user can raise an additional issue which has not been conceived by the system at all. In order to do this, the new issue has first to be named, and then the kind of data from the MIKE SHE model which will be used for judging the new issue have to be chosen. The user can choose from the following options:

-Water content in the unsaturated zone
-Maps of depth to phreatic surface
-Time series of depth to phreatic surface

After choosing the data type, the individual user can choose his or her location of interest through the map interface. Once the results are received from the fact engine, the individual user has to provide the issue judgements for the new issue in linguistic terms. Through this procedure, the individual user will formulate a decision matrix with 5 columns (4 institutional issues + one own issue; see Table 4.6).

Table 4.6. Example of optional decision matrix for Individual User

	Issues				
	Citadel	*Forest*	*Corn fields*	*Water price*	*Own issue*
Scenario	Provided by LA (linguistic)	Provided by EA (linguistic)	Provided by FC (linguistic)	provided by WS (crisp; euro/ m^3)	Provided by IU (linguistic)
Scenario 0	Fairly bad	Good	Good	0.5	Good
Scenario 1	Fairly good	Extremely bad	Excellent	0.55	Bad
Scenario 2	Very bad	Good	Medium	0.60	Fairly bad
Scenario 3	Good	Excellent	Extremely bad	0.45	Very Good
Scenario 4	Bad	Bad	Very good	0.70	Medium
Scenario 5	Fairly bad	Fairly good	Very bad	0.55	Fairly good
Scenario 6	Very good	Good...	Very bad	0.65	Good
Scenario 7	Very good	Fairly bad	Bad	0.75	Medium

Regardless of their role, users must first implement the judgement stage. After the judgement stage they will all have a properly-formulated decision matrix where most of the issue judgements are common for every player/participant. In the second, *evaluation,* stage the users provide the evaluation data for the issues at hand. Here every user is truly independent. The first step in this stage is the classification of the issues as costs or benefits. In principle it appears as though this step is very obvious. From the four main issues, all of them are benefits except water price, which is a cost[14]. However, the users are given the freedom to

[14] In the economic model of the Water Supply Company water price enters as benefit and higher water prices bring higher profits. However, high water prices will not be acceptable to the community. The Water Supply Company will actually try to maximise its profits with reasonable (acceptable water price). Here the issue of water price is probably best described by non-monotonic utility function (increasing and then decreasing), but

declare any issue as a cost or as a benefit. The cost – benefit codes will then be inconsistent over different players, but for the individual ranking of the scenarios this is irrelevant. The advantage of this situation is that it allows the users to declare *individually* a certain issue as a 'cost' even if it is accepted as a 'benefit' at a social level. This behaviour might be damaging to the whole process, because it allows for individual evaluations which are the *reverse* to those common for the community (declaring the issue 'Cornfields' as a cost, for example, means that the *worse* outcomes for this issue are better for that individual). However, all judgements and evaluations in *Aquavoice* are made public through the system, and such steps will be visible to all other participants. On the other hand, the behaviour of some institutional player during the course of the game may be such that it drives the other players to 'punish' that player through such downplaying of issues of value to that institution.

The second step in the evaluation stage is the provision of the preference weights for the issues. They are provided as points between 0 and 100 (converted later into a scale between 0 and 1). In the *Aquavoice* context, the assignment of preference weights may be understood as an answer to the question of how much that individual user 'cares' for a given issue as compared to the other issues.

After all data on judgements and evaluations are provided, the adapted FMADM algorithm calculates the individual rankings of the proposed scenarios. As already introduced in the description of the method, the individual ranking is calculated on the basis of the relative closeness of the scenarios to the ideal solution. In this arrangement each user has a different ideal solution. The results obtained in terms of relative closeness to the ideal solution will therefore be meaningless to other players/participants. In order to present the results to all users in a consistent way, using comparable scales, two different normalisation approaches are used. The first one is the simple, so-called, 'Borda score' where, for each of the 8 scenarios, scores of 8,7,.....,2,1 are assigned to the first ranked, second ranked,...,last ranked scenario. This approach does not preserve the information about the mutual closeness between the scenarios.

The second normalisation uses a linear scale transformation, where the relative closeness for each scenario is divided by the scenario with maximum relative closeness. This second normalisation ranks the scenarios on a normalised scale between 0 and 1 with the top scenario always receiving a score of 1. This score is called the 'utility' score and it still gives the possibility to see the mutual closeness between scenarios. As we shall see later, both scoring scales can then be used for the aggregation of the evaluation results of all users in order to present the overall response of the whole community to the proposed scenarios. Examples of the Borda and the 'utility' scores obtained as results from the individual ranking are presented in Figure 4.32.

---

since the method requires monotonic utilities in this evaluation even the Water Supply Company needs to consider water price as a cost.

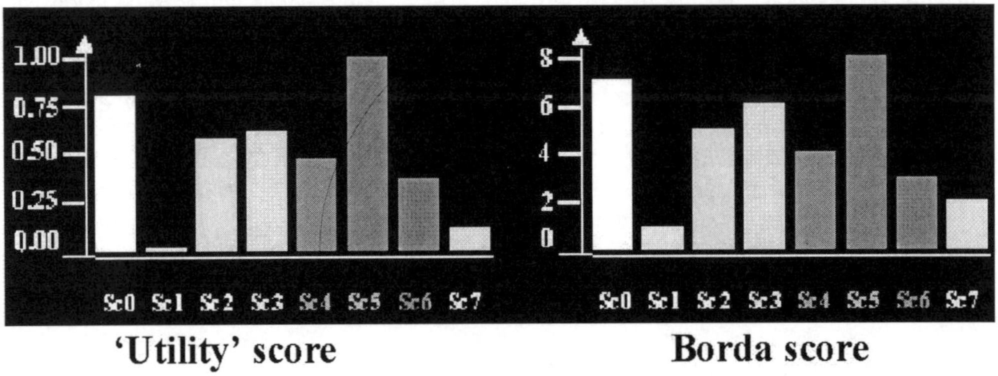

Figure 4.32. Examples of Borda and 'utility' scores obtained as a result of individual ranking of scenarios

With these results of the individual scenario rankings, the actions of the players/participants in the judgement engine component are completed. On the basis of these results, an *Aquavoice* session can proceed further with the processes of negotiation and collaboration. It needs to be realised, however, that the whole process is not at all linear, and that at any moment in time any user can update his or her judgement or evaluation. This may come as a result of some negotiated agreement, or because of some change in attitudes that will shift the position of any player/participant. In any case, the updated judgements or evaluations are immediately reflected through the system to all other players/participants.

4.2.3 Negotiation and Collaboration Platform

In order to make the processes of negotiation and collaboration more dynamic and more interesting, many of the additional features of *Aquavoice* are excluded from their direct treatment in the judgement and evaluation procedures and they do not enter any of the decision matrices directly. These additional features are mainly related to the four institutional roles.

The Local Authority, for example has at its disposal a certain amount of tax-generated revenue. An implicit goal of the Local Authority is to keep this amount as much as possible for other purposes which are not related to the current project. On the other hand, because the Local Authority is very much in favour of the project, it can decide to spend parts of this revenue for direct support of the project. One part can be invested for protecting the Citadel region if the rest of the community prefers an unfavourable scenario for the Citadel and the Local Authority will be otherwise 'forced' to accept it. Another part can be offered as compensation to the Farmers Council in order to persuade them to accept some scenario which is perceived as unfavourable by the farming community. The amount offered as compensation is an important negotiation issue between these two players/participants.

The Environmental Agency has the powerful role of setting time limits to the abstraction licence for the Water Supply Company. This will probably be one of the central negotiation issues between these two parties. The Water Supply Company, on the other hand, although aiming at maximising its profit, will sometimes need to spend parts of this profit to help in the persuasion of the Farmers Council to accept a certain scenario. With such a position of the players, while assuming that all participating users are aiming at reaching some agreement that will give a green light to the project, the roles of the Local Authority and Water Supply Company are particularly interested in achieving a positive result.

There are no such features pre-defined for the role of individual users. It is expected that with larger numbers of individual users participating, new alliances may emerge, possibly even around new issues of concern, previously not defined. Their reaction to the proposed scenarios can very much influence the outcome of the negotiations among the four institutional role players.

The platform for negotiation and collaboration itself currently provides only basic support tool for these processes. It is the least developed component of the system and lacks many of the design requirements as these were stated in Chapter 3. Nevertheless, in order to satisfy at least the basic needs for negotiation and collaboration among the players/participants, it has been designed with two requirements in mind. Firstly, in order to have some representation of the 'social landscape' in terms of the overall response of the whole community to the proposed scenarios, the results of the individual rankings have been aggregated. Secondly, all changes in judgements and evaluations, while being automatically updated, are made *transparent* to all the participants in the system.

For the first requirement, a very simple aggregation procedure has been chosen. All the scores of the individual rankings of the scenarios are simply added to represent the overall score result representing the response of the whole community. Both the Borda and the 'utility' scores are treated the same way. An example for the Borda score is given in Figure 4.33.

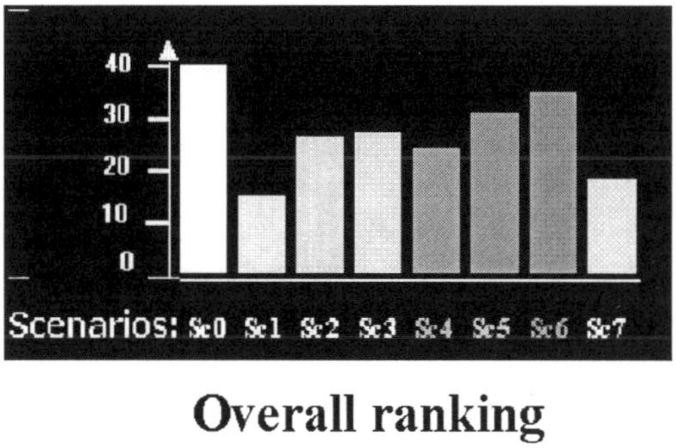

Figure 4.33. Overall Borda scores for all scenarios

Regardless of the role, each individual ranking score enters with equal weight in the summation. In other words, the rankings of the institutional roles have the same influence as the rankings of any individual user. With large numbers of individual users participating, this means that their influence will be dominant in the formation of the overall response. In some sense, this counter-effects the dominance and control over the main issue judgements otherwise apportioned to the institutional roles.

This simple aggregation procedure has been chosen mainly because it should serve only as a starting point for further negotiation and collaboration and no attempt has been made here to provide some objectively 'fair', let alone 'just', mapping from individual preferences into a social preference. As we have already mentioned, there are many methods which try to objectivise this process, and many of them are topics in the field of 'Group Decision Making under Multiple Criteria'. In fact one of the existing methods proposes the combination of the TOPSIS method carried out independently by every individual and the summation of the Borda scores of the individual rankings for obtaining the overall social preference (see Hwang and Lin, 1987). Essentially, the same procedure is being followed here, but with many more individual 'decision makers' being part of the group, and without attempting to recommend the result obtained as final. Instead it is only an indicator for the overall 'feeling' of the whole community.

The second aspect of the negotiation platform, related to the transparency of all the results and all the actions taken by the players/participants in the system, is closely linked to the overall design of *Aquavoice*. It will become much clearer with the presentation in the following section that is devoted to the implementation architecture of *Aquavoice* and the dynamics of the play. It may suffice to say here that the platform is designed in such a way that at all times the user can have access to his or her own evaluation results, to the results of the four institutional role players, and to the overall ranking results. This is achieved through updating the web pages automatically with all the information pertaining to the judgement and evaluation of any user. The actual negotiations are carried out through a chat-like interface embedded in one of the main pages. Through this interface the judgement and evaluation data of any user participating in the discussion is accessible.

4.3. The Dynamics of the *Aquavoice* game

The dynamics of the *Aquavoice* game are best described by first introducing the more detailed implementation architecture of the system, and especially that part which enables the integration of all the activities taken by each role player involved. Although the overall implementation architecture has already been introduced in the previous section, this present section which deals with the *Aquavoice* dynamics is a more appropriate place for the introduction of its detailed architecture. The steps taken by each role player through his or her respective user interface, and the ways in which the individual actions are being integrated into one collaborative framework, will become clearer by following them within the detailed implementation architecture. As presented in Figure 4.34, the detailed implementation architecture follows again the main three-tier approach.

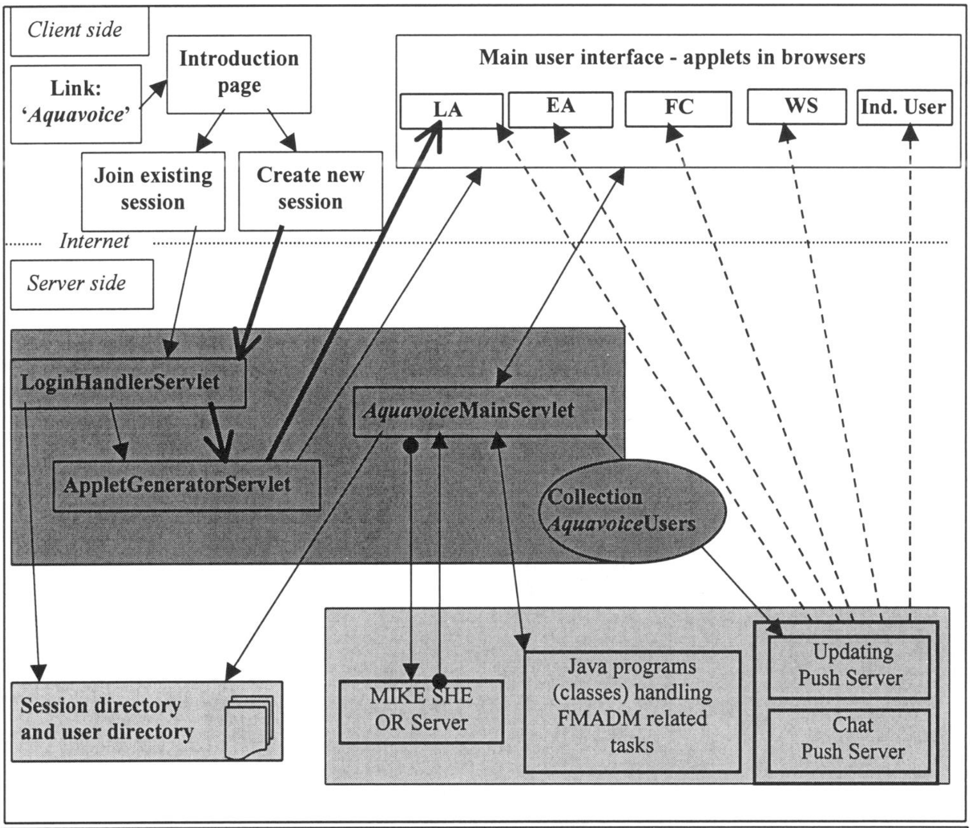

Figure 4.34. Detailed implementation architecture of *Aquavoice*

Let us begin with the procedure for initiating an *Aquavoice* session and how the architecture supports it. An *Aquavoice* session is designed to last for a limited time period. The duration of the session is envisaged to be between several hours and one month, and it will be determined by the initiator of the session, which in all cases is the Local Authority role player. Short sessions are envisaged to be played intensively when all players/participants are continuously logged into the system, while longer sessions can be played with on-and-off participation. Only after the initiation of a session by the Local Authority, can all other players join that session. At the moment, the system is configured in such a way that it can support 5 parallel and independent *Aquavoice* sessions.

The *Aquavoice* game is accessed by following the link '*Aquavoice*' from the second entrance page into the system (Figure 4.34; see again Figure 4.5 as well.). The first page (Figure 4.35) gives an introduction to the game and a brief description of each role. At the bottom of the page two options are presented: the user can either join an existing session or create a new session.

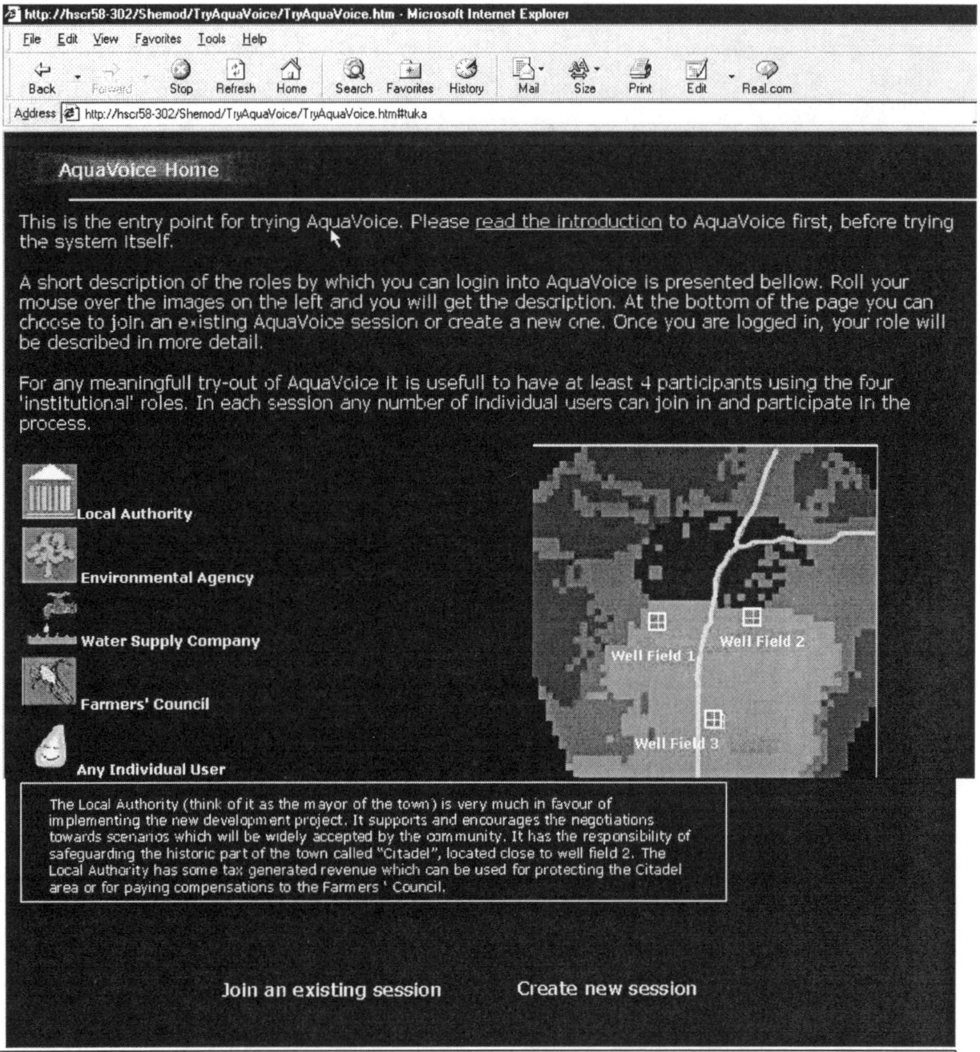

Figure 4.35. Introduction page with two options: create a new session or join an existing one

Choosing to create a new session means that the user will necessarily take the role of the Local Authority. Along with providing a user name and a password, this role player sets the time duration for the session. For an easy identification of the session, its name is the same as the user name of the Local Authority role player. The interface for creating a new session is actually used only once. In subsequent logins, the Local Authority role player uses the interface for joining an existing session like all the other players/participants.

Choosing to join an existing session leads the user to a page where all active sessions are listed with details about their duration and about the roles that are taken. The user can choose one of the active sessions, and by providing a user name and a password he or she can join the session. Since for an existing session the role of the Local Authority is already taken, the user can either take one of the three remaining institutional roles, or, if they are already taken as well, he or she can join the session with the role of an Individual User. Any user that has already joined some session, and later logged off (in cases where the game lasts for a longer time period), can join that session next time through the same interface. Figures 4.36 and 4.37 show the user interfaces for both options. The actual login into the system happens through one of these two interfaces.

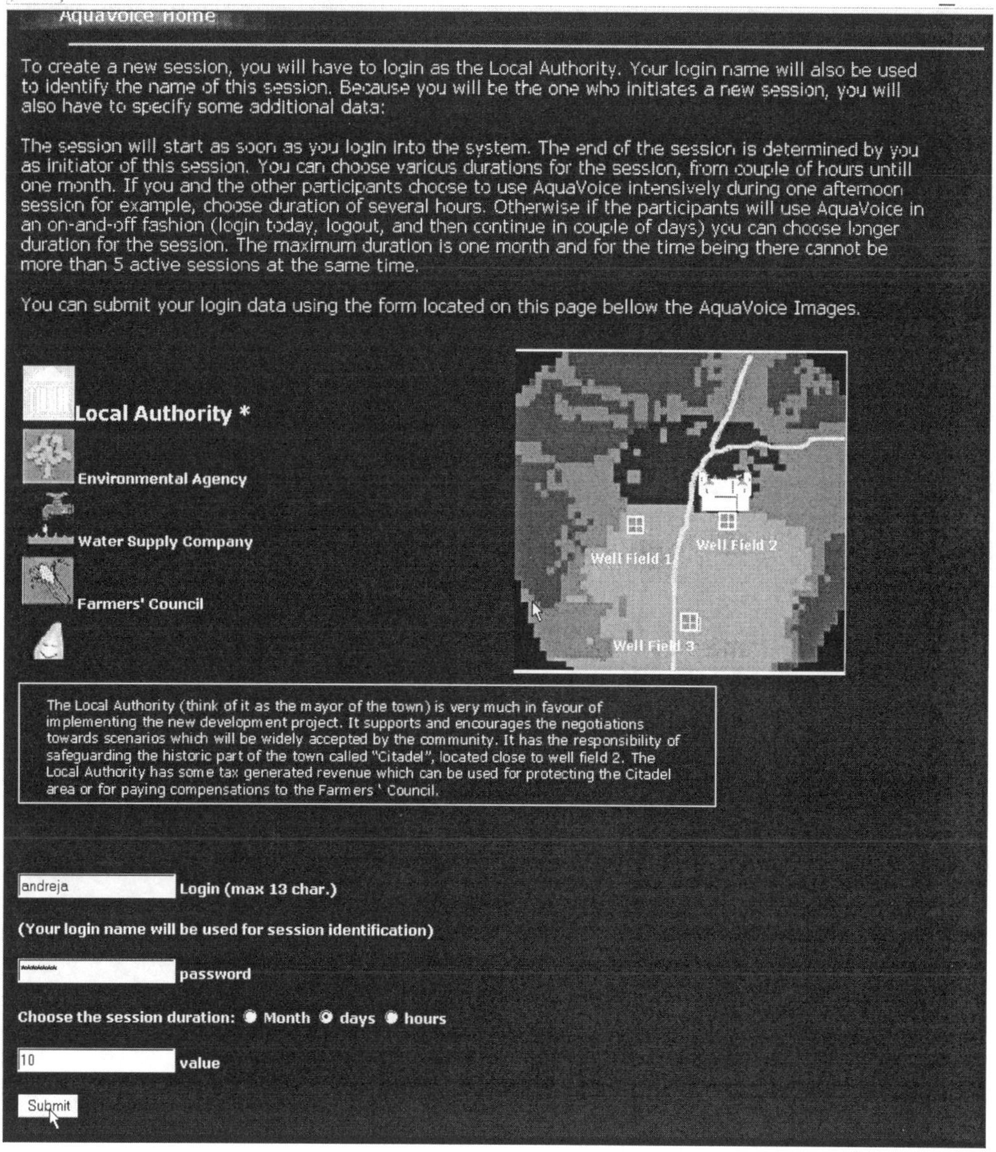

Figure 4.36. Interface for creating a new session (Local Authority role)

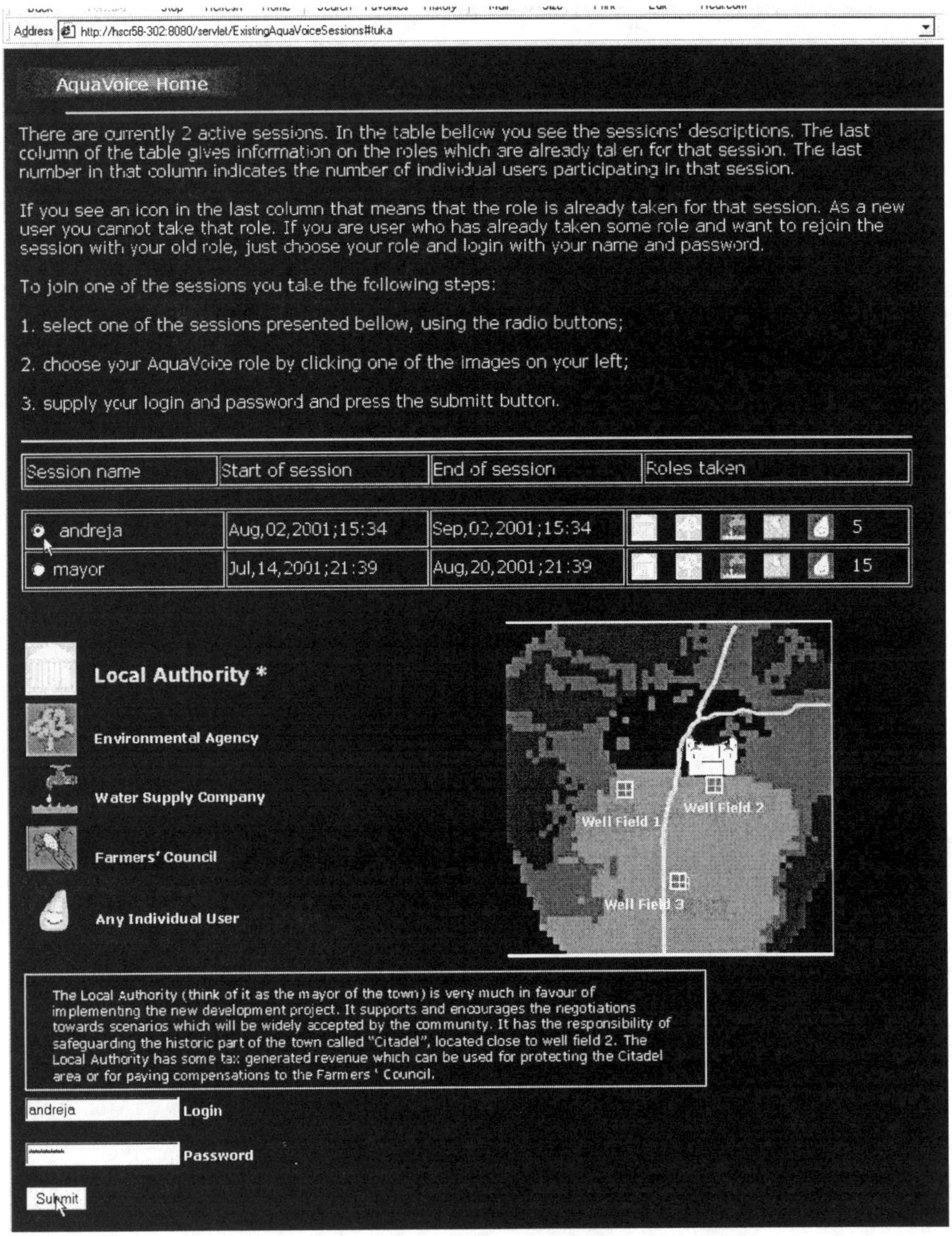

Figure 4.37. Interface for joining an existing session. (Any role)

The control and management of *Aquavoice* sessions is carried out by a number of servlet programs located in the middle tier (Figure 4.34). The first servlet named *LoginHandlerServlet* handles all the login data. When a new session is initiated it creates a separate folder in the back end of the server where all the data about the new session are stored, and prepares all necessary files with default data. When a new user joins an existing session, the servlet adds this user in to the session directory and creates the necessary workspace for that user (the user directory).

If an existing user joins the session the servlet reads the stored data for that user from the previous login and allows this user to join the session with the old 'state' of that user. Apart from the login name, the password, the role taken, and other identification data, the system stores all the data related to users, judgements, evaluations, individual rankings of scenarios, etc. Upon login, all these data are updated with all changes that have occurred in the time that that user was absent from the session. This last task is very important because in the meantime many changes may have occurred that influence directly the 'state' of that user. (For example the issue judgements by the institutional players may have changed). The updating is achieved in such a way that, for all registered users, regardless of whether they are on-line or not, the state is continuously updated. Those who are currently on-line will be able to see the updates immediately, while for those who are not online the information is stored on disk in their respective user directories. Upon their next login, the updated state is read from disk and they join the session in its current state.

Once the login procedure is completed, the *LoginHandlerServlet* initiates a procedure for sending back the page with the main interface for the role-player. This procedure invokes another servlet program, called *AppletGeneratorServlet,* that, according to the session, user name, and the role taken, sends the web page with the appropriate applet to the user's browser. This web interface contains the main applet in its upper frame, and a frame reserved for help pages in its lower frame. When the *AppletGeneratorServlet* sends the page to the user it also gives the applet the information about which *AquavoiceMainServlet* is going to serve this user. Each session has its own *AquavoiceMainServlet*, which serves all the users in that session. All the activities of the users are handled through this servlet. When a request comes for obtaining some data from the fact engine, it initiates the communication with the MIKE SHE OR (Output Retrieval) server, in a similar manner as the Servlet_OR in Figure 4.14. When a user submits all the data about his or her judgements or evaluations, the same servlet invokes the additional Java programs that run the FMADM algorithm (conversion of the linguistic terms used for the issue judgements, and running the TOPSIS algorithm).

Apart from all the judgement and evaluation handling the *AquavoiceMainServlet* is responsible for all the updating tasks. Whenever a user makes a new judgement or a new evaluation, the 'state' of that user and, if necessary, of all other users, is updated. This updating is done in three ways. Firstly, all information is updated and stored in all respective user directories. Secondly, all the dynamically created web pages, which contain this information and are used for a transparent presentation of one user's position to all the others are updated and stored on the web-server. Thirdly, the same updating information is held in a special collection (a Java object) of users called *AquavoiceUsers*. This collection holds each user as an object with its own judgement, evaluation and ranking data. The *AquavoiceUsers* collection is very important because it is used for the instantaneous updating and sending of data to all users who are currently on-line. *AquavoiceMainServlet* holds this object always in memory. If the servlet engine stops, this object is saved on disc in such a way that its current 'state' is preserved, and can be reloaded upon the next start-up.

The state of the *AquavoiceUsers* collection object is controlled by the *AquavoiceMainServlet*, and is being changed by the actions of the users. The same object is also shared by the *Updating Push Server* implemented on the server side, which is responsible for sending the updated data to the users that are currently on-line. This server is started and controlled by the *AquavoiceMainServlet.* Whenever changes occur, the updated state of the *AquavoiceUsers* collection is read by the push server and sent to all users. The use of push servers for this purpose was chosen in order to have immediate updates of all the data. As has been already mentioned, the use of servlets for this task is less appropriate because they are 'request-response' ('pull'-orientated) programs.

Another push server that is also started and controlled by the *AquavoiceMainServlet* is the *Chat Push Server,* which enables the negotiation among the players through their chat like interface. Both the *Chat Push Server* and the *Updating Push Server* use the multithreading approach for communicating with the clients.

This description of the implementation architecture of *Aquavoice* still excludes many smaller programs that are responsible for other supporting tasks of the system. For example, a separate servlet always reports the time at the *Aquavoice* server so that all users will have the same information about the duration of the session and the time expired. The servlets invoke many smaller utility programs that handle the proper storage and retrieval of all the files at the back-end server and on the web server, perform all necessary checks, and report eventual error messages to the users. At the front-end, the user interfaces are enhanced by making use of JavaScript functions that are linked with the main Java applets.

In what follows we will concentrate only on the main user interface features for the five respective roles. By going through all the steps taken by each role player, the 'feel' for the dynamics of an *Aquavoice* session will hopefully become closer.

Local Authority

We will begin by introducing the steps taken by the Local Authority. For this role we will introduce all the steps, but, because many steps are similar for all the roles, for the rest of the roles we will concentrate only on those steps that are different. The main applet has a wizard-like interface through which the user can go from one step to another by using 'Previous' and 'Next' buttons. This same approach is taken for the design of the interface for all roles.

Step1 (Figure 4.38)

After login, the two-frame web page is delivered to the user, with the main applet in the upper frame, and the help pages in the lower frame. The first step of the Local Authority role is to choose what kind of MIKE SHE model output data will be used in his or her judgement about the Citadel issue (Figure 4.38)

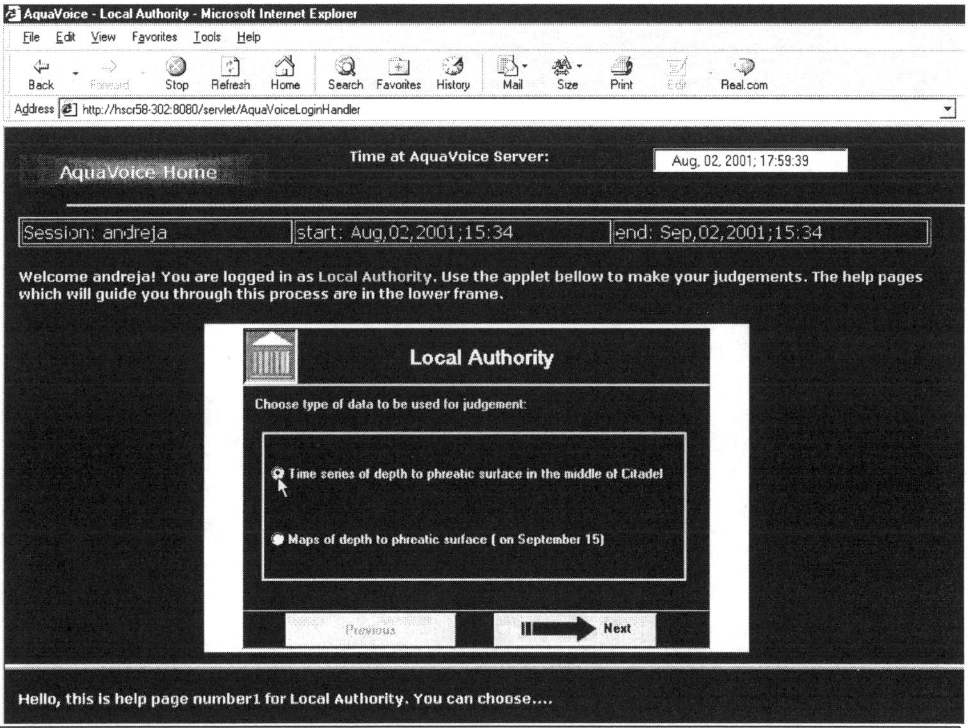

Figure 4.38. Step 1 (LA). Choosing kind of data for judgement of the Citadel issue

Step2 (Figures 4.39a and 4.39b)

In this step the chosen data are obtained from the server (Figure 4.39a). After obtaining the chosen data from the fact engine, a separate frame presents the user with his or her latest judgement data from the user's most recent participation in the session. The main judgement data in this example are issue judgements about the issue Citadel. Data about the readiness to spend financial resources for the protection of the Citadel region and for compensations to the Farmers Council is also presented (Figure 4.39b). If during the most recent participation the user provided some textual explanation of his or her judgement, it is also shown in the interface. If this is the first login of the user, none of these data is yet set.

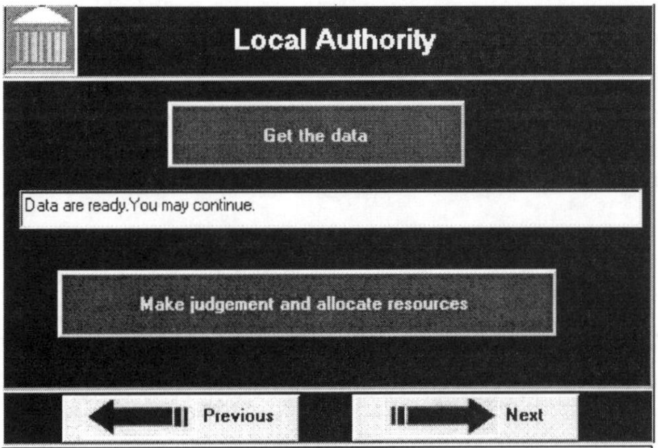

Figure 4.39a. Step 2 (LA) Getting the data

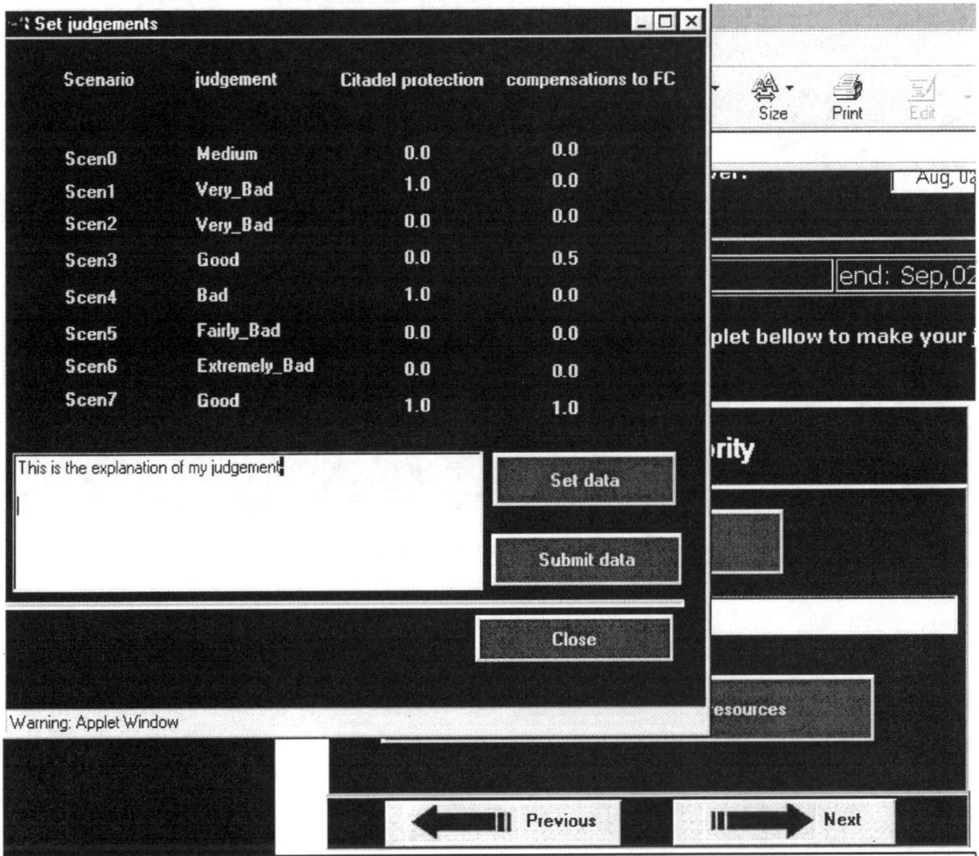

Figure 4.39.b. Step 2 (LA). Viewing the existing judgement data

Step 3 (Figure 4.40)

New judgement input is enabled through another frame that contains the main interface for viewing the MIKE SHE outputs. The example shown on Figure 4.40 is with time series of depth to phreatic surface. The results for all scenarios are presented on one graph in different colours. For each scenario the role player looks at the graph and provides an issue judgement in linguistic terms and sets the amount of financial resources that may be spent for the protection of the Citadel region or offered to the Farmers Council as compensation.

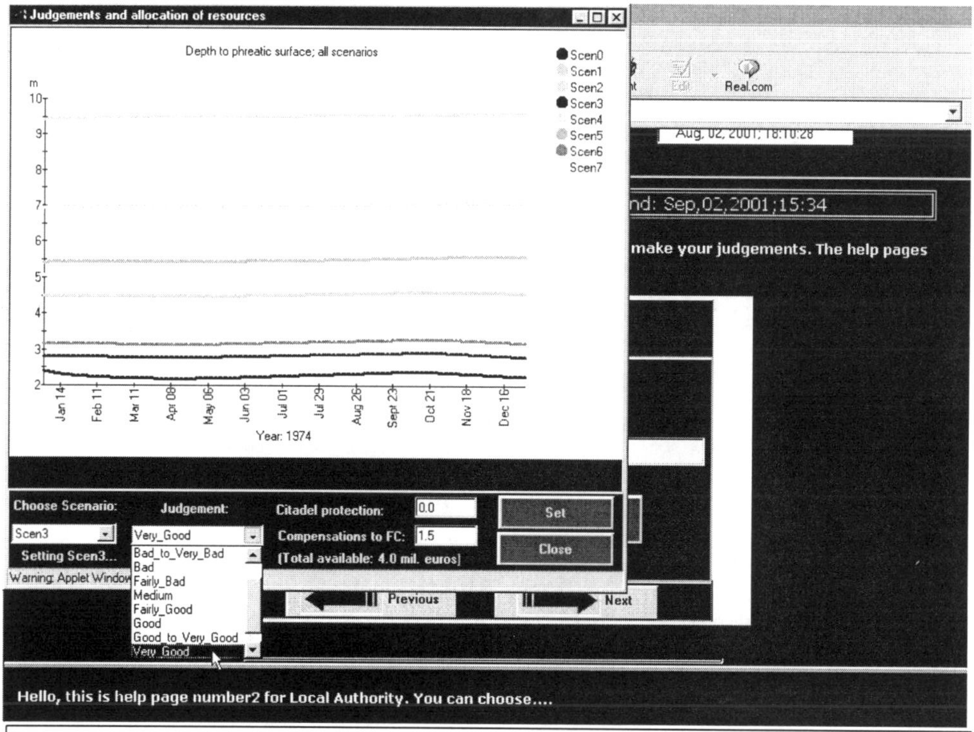

Figure 4.40. Step 3 (LA). Input of new judgement data

Step 4 (Figure 4.41)

After providing new data, the role player can give a new explanation of his or her judgements using the text box in the underlying frame. Once all the data are ready, all judgement data are sent to the server, where they are stored and the user's web page is created (Figure 4.41)

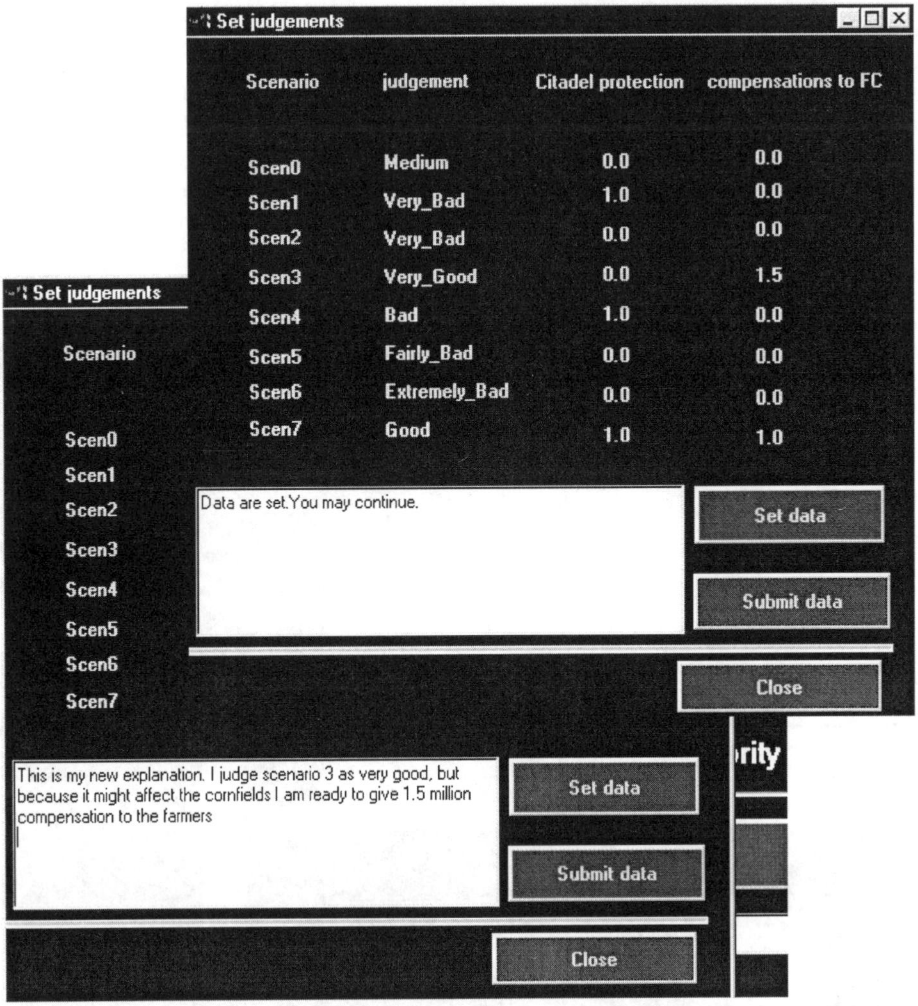

Figure 4.41. Step 4 (LA). Input of judgement explanation and submitting the data to the server

Step 5 (Figure 4.42a and 4.42b)

The user can check the created web pages for his or her own issue, or for the issues of any other institutional role.

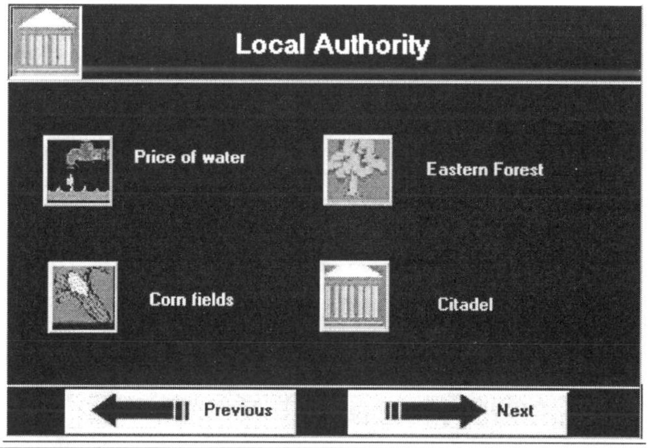

Figure 4.42a. Step 5 (LA) Links to the web –pages for the four institutional issues (and role players)

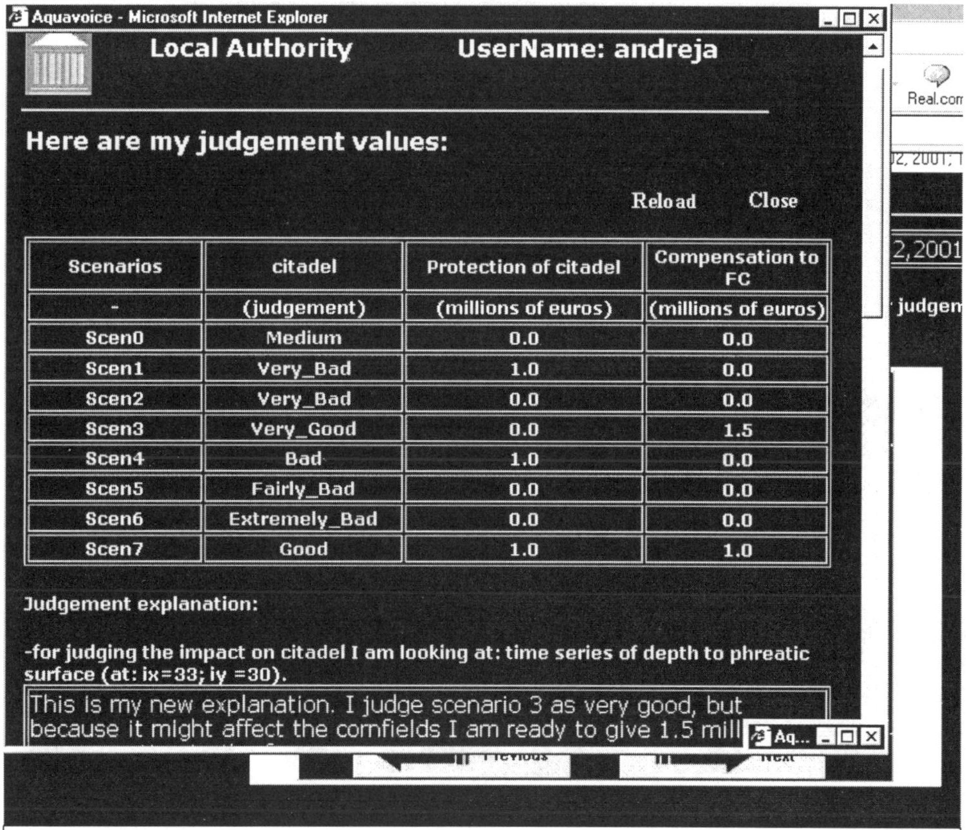

Figure 4.42b. Step 5 (LA) Example web-page with judgement data of Local Authority

Step 6 (Figure 4.43)

The user moves to the evaluation stage. The role player indicates which issues are costs an which are benefits, and assigns the preference weights for the four issues.

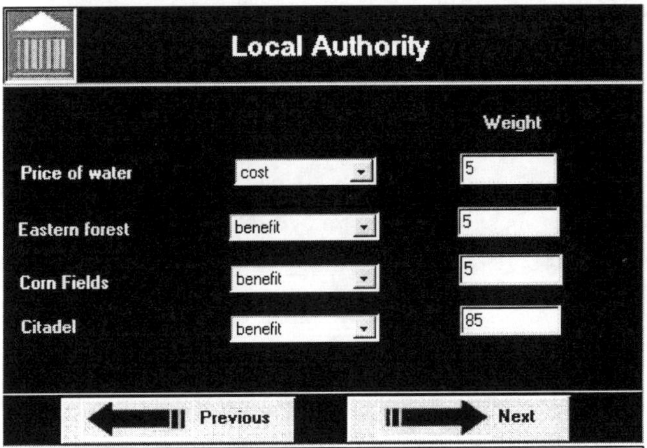

Figure 4.43. Step 6 (LA) Input of the evaluation data

Step 7 (Figure 4.44a and 4.44b)

The user provides textual explanation of his or her evaluation and sends the evaluation data to the server (Figure 4.44a). Upon completion of the evaluation, the web page is updated with the evaluation data and the results for the individual rankings of the scenarios Figure 4.44b).

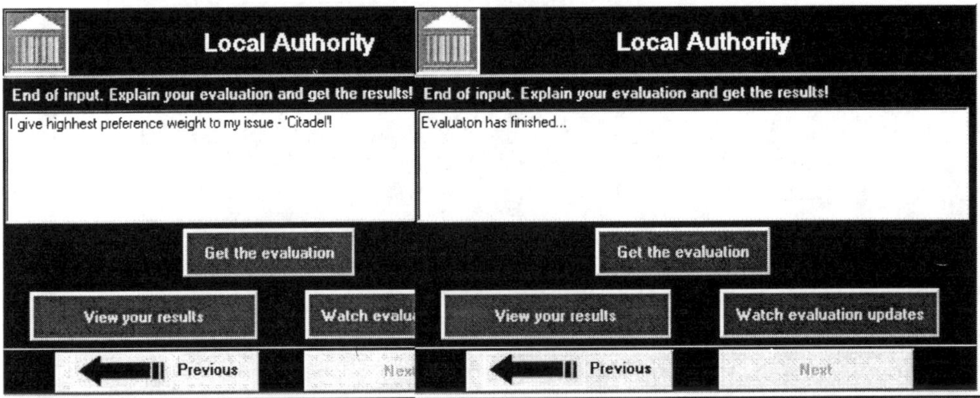

Figure 4.44a. Step 7 (LA) Textual evaluation explanation and submitting the evaluation data to the server

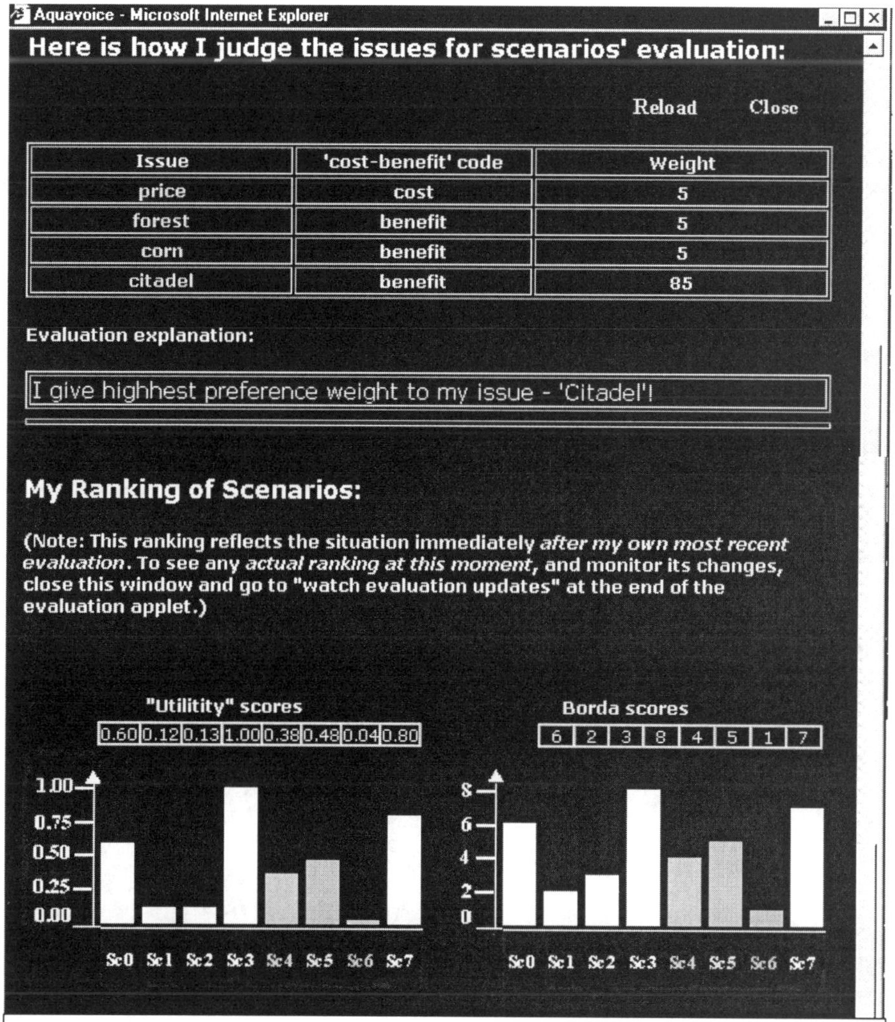

Figure 4.44b. Step 7 (LA) Viewing the evaluation data and the individual ranking of scenarios

Step 8 (Figure 4.45)

By following the link 'Watch evaluation updates' the user loads a separate web page, which is connected to the *Updating Push Server* and the *Chat Push Server*. The page shows the rankings of all four main institutional players, the ranking of the actual role player who is viewing the page (in this example the Local Authority role player) and the overall ranking of the whole community. Whenever some user makes a new judgement or evaluation, the results are automatically sent to this page and the charts are updated. By clicking on the icons representing the four institutional players, their web pages are accessible directly. The bottom of the page contains a text box where all actions (changes in judgements or evaluations) by any on-line user are reported.

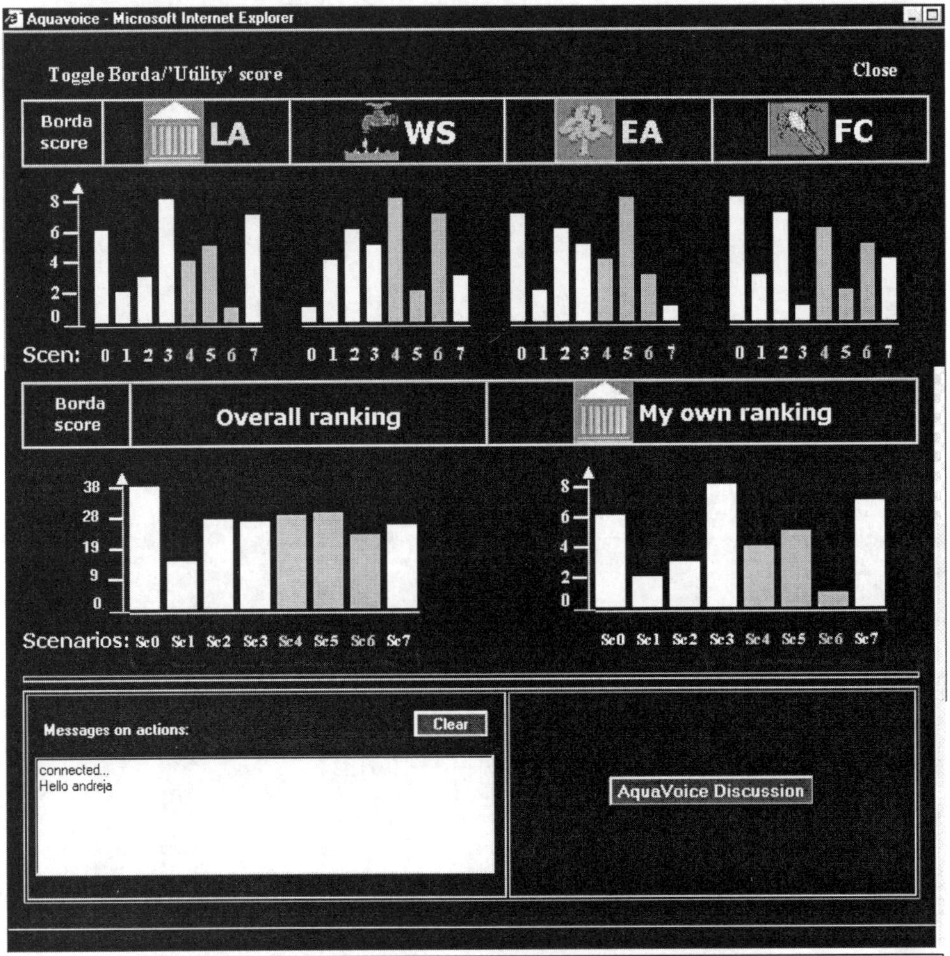

Figure 4.45. Step 8 (LA) Web page for watching evaluation updates

Step 9. (Figure 4.46)

By following the '*Aquavoice* Discussion' link the role player can initiate or join the negotiations with other players. The user is automatically identified and logged into the discussion session. Through this interface the web pages of all users who are currently in the discussion, *regardless of their role*, are accessible (This is achieved by typing 'url/username' in the discussion field of the chat interface). This provides the required transparency of each user's position.

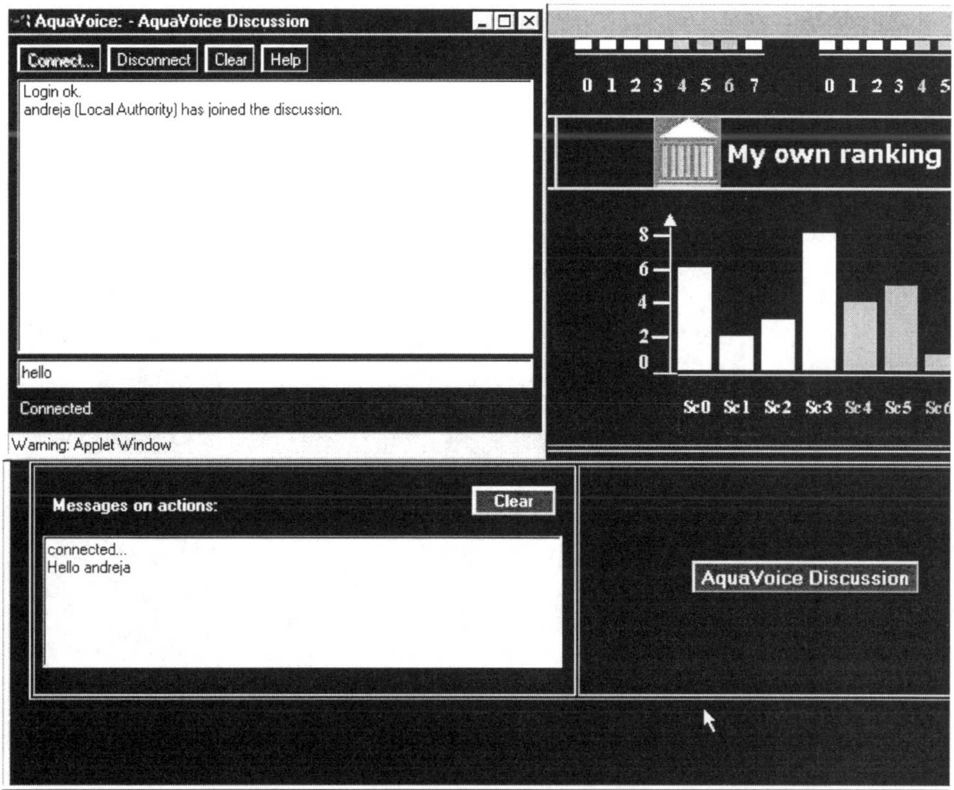

Figure 4.46. Step 9 (LA) Participation in negotiations through a chat-like interface

As has already been mentioned, many of the steps are similar for all role players. Therefore, in presenting the steps for the other four role players, we will concentrate only on those steps that are specific for a given role player.

Environmental Agency

Step 1 (Figure 4.47

Similarly to the Local authority, in the first step after login the Environmental Agency chooses the kind of data that will be used for making its judgements. In this example, these data are maps of depth to phreatic surface.

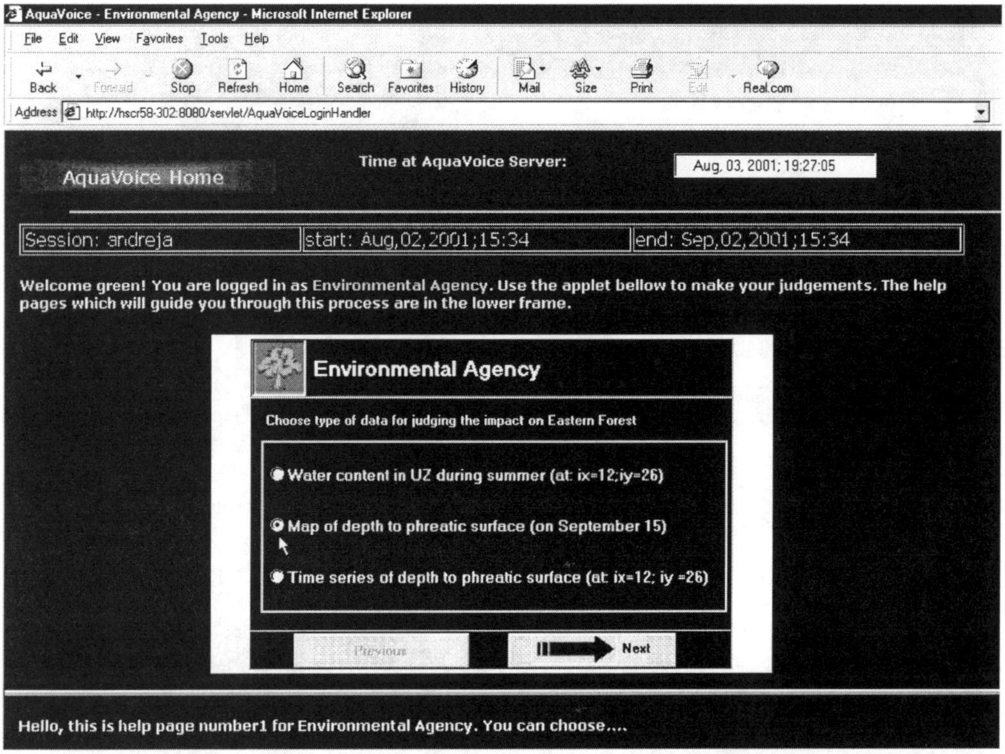

Figure 4.47 Step 1 (EA) Choosing kind of data for judgement of the Forest issue

Step 2 (Figure 4.48

This step is again similar to the equivalent one of the Local Authority, only that now the Environmental Agency uses judgements for the influence on the Forest, and sets the duration for the abstraction licence (Figure 4.48.

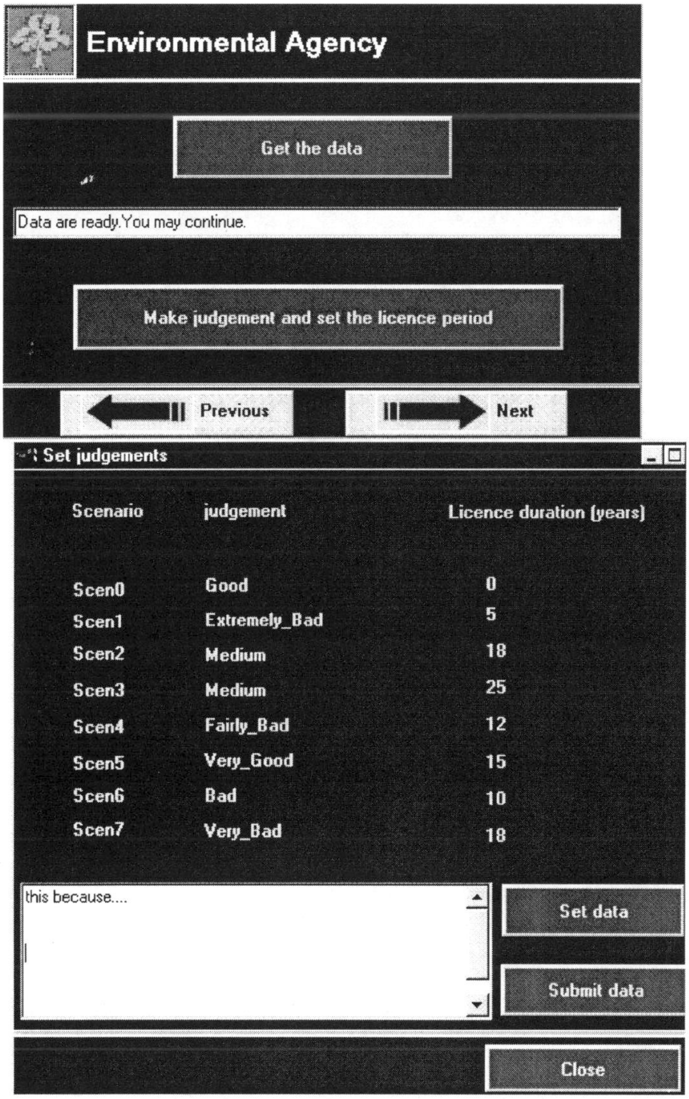

Figure 4.48 Step 2 (LA) Getting the MIKE SHE model outputs and viewing the existing judgement data

Step 3 (Figure 4.49)

In this step the Environmental Agency role player looks at the maps of depth to phreatic surface for all scenarios and provides the new issue judgements. For each scenario it sets the duration of the abstraction licence (Figure 4.49.

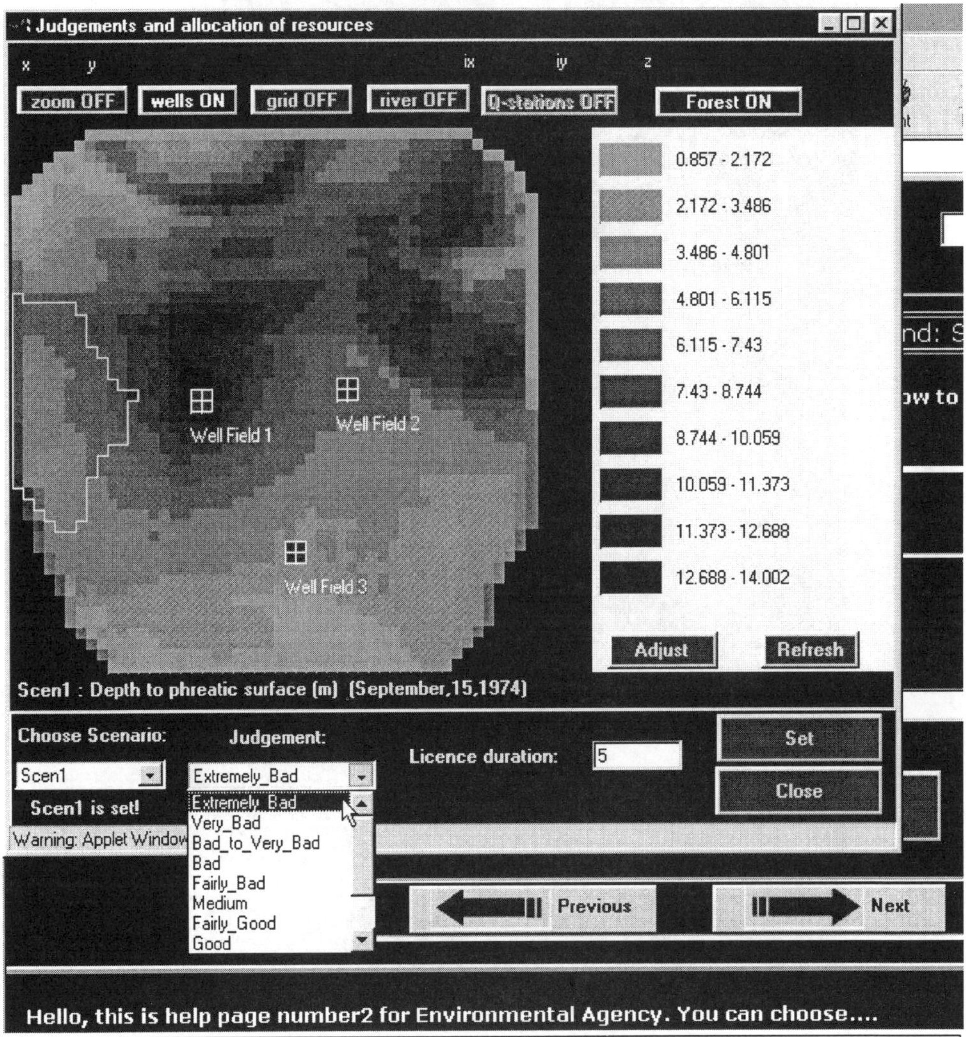

Figure 4.49 Step 3 (EA) Input of new judgement data

Since the subsequent steps of the Environmental Agency role player are same as for the Local Authority, their presentation can be omitted.

Farmers Council

Step1 (Figure 4.50)

This role player chooses the kind of data that will be used for making its judgements. In this example, the values of the water content in the unsaturated zone during the summer period are chosen.

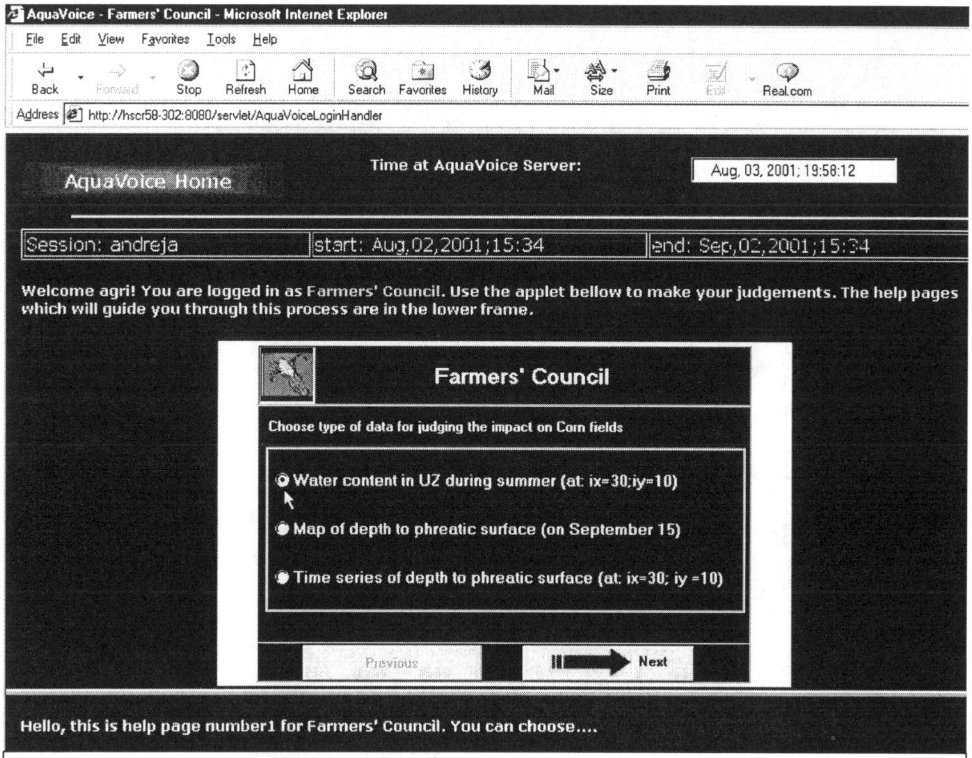

Figure 4.50 Step 1 (FC) Choosing kind of data for judgement of the Corn Fields issue

Step 2 (Figure 4.51)

The role player gets the requested results and looks at any existing judgement data. The issue judgements about the Corn Fields issue are in linguistic terms, and the demands for compensations are in millions of euros. (Figure 4.51).

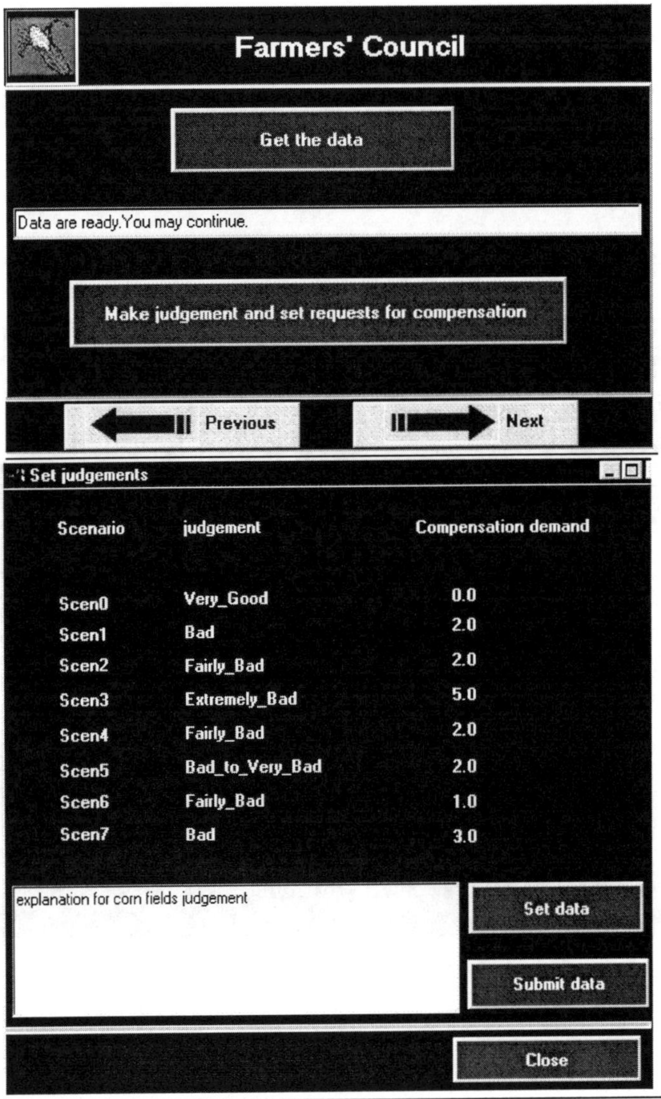

Figure 4.51 Step 2 (FC) Getting the MIKE SHE model outputs and viewing the existing judgement data

Step 3 (Figure 4.52)

In this step the Farmers Council role player looks at the graphs of water content distribution in time and depth for all scenarios and provides the new issue judgements. For some scenarios the FC will require a certain amount of resources as compensation for damages (Figure 4.52).

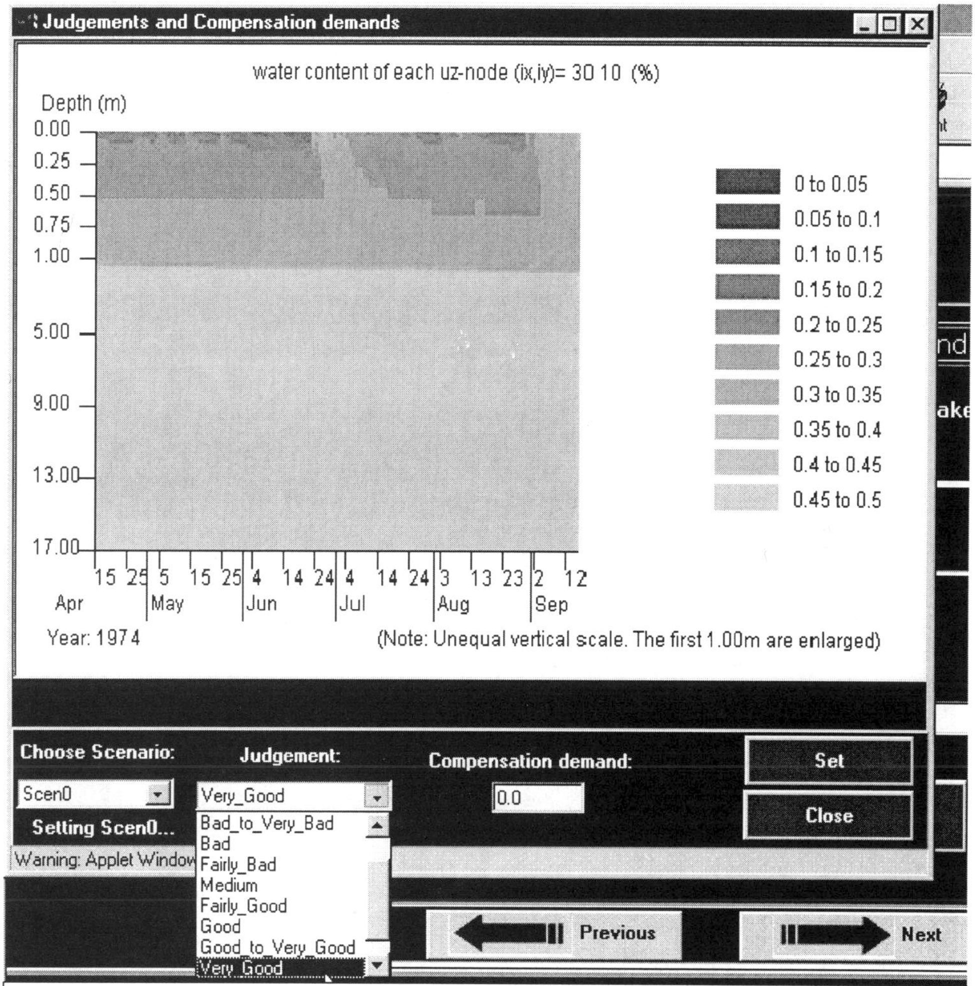

Figure 4.52 Step 3 (FC) Input of new judgement data

Again, the remaining steps are similar to the ones taken by the Local Authority and will not be presented.

Water Supply Company

Step 1 (Figure 4.53)

The role player with the role of Water Supply Company has to take different steps in the judgement stage. This role player does not use the MIKE SHE model outputs, but uses its own economic model for calculating water prices. In the first step, the player obtains the durations of the abstraction licence for the 8 different scenarios.

In case the Environmental Agency has not yet joined this session, then the duration data provided will be default data set by the system. The example shows the case when the role player obtains the duration data as set by the Environmental Agency (Figure 4.53). Only after these data are obtained can the player go to the second step, in which the water prices will be set.

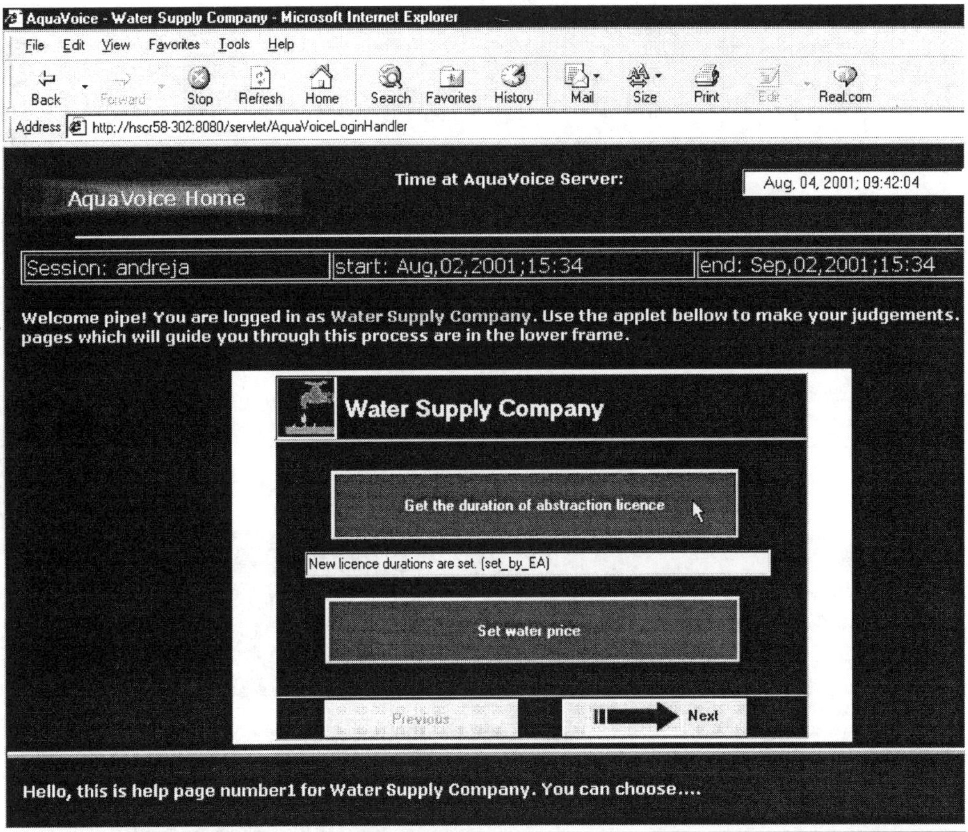

Figure 4.53. Step 1 (WS) Getting the duration of abstraction licence

Step 2 (Figure 4.54a and Figure 4.54b)

In this step, similarly to the situation described for the other roles, old judgement data from the most recent login are presented (Figure 4.54a). The data are in terms of water prices and profits for each scenario. The user then chooses the scenario for which he or she wants to change some of these data by using different inputs in the economic model. The text field is again used for providing textual explanations of the judgement data. Before submitting the data, the role player can choose whether the profit data are going to be visible to the other players/participants or not.

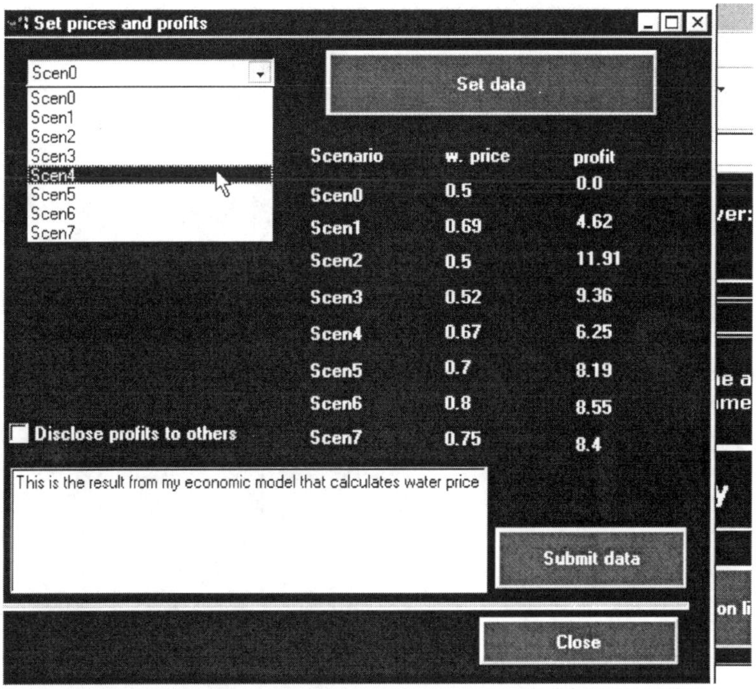

Figure 4.54a Step 2 (WS) Viewing the existing judgement data and choosing a scenario for which new data will be set

The interface for using the economic model is shown on Figure 4.54b. On top of the frame, the abstraction data for the chosen scenario are presented together with the duration of the licence period. On the left side, two charts show the economic performance of the scenario. The upper chart plots the cost and benefit curves in time. The lower chart plots the profit (NPV) curve. From these charts it is easy to see when the initial costs are recovered and thus when the scenario that is chosen starts earning profits. On the right side of the interface the role player can set different water prices and obtain different results for profits. The graphs are updated with the new data. Apart from the water price, the role player can choose, if necessary, to sacrifice some of the profits to offer compensations to the Farmers Council.

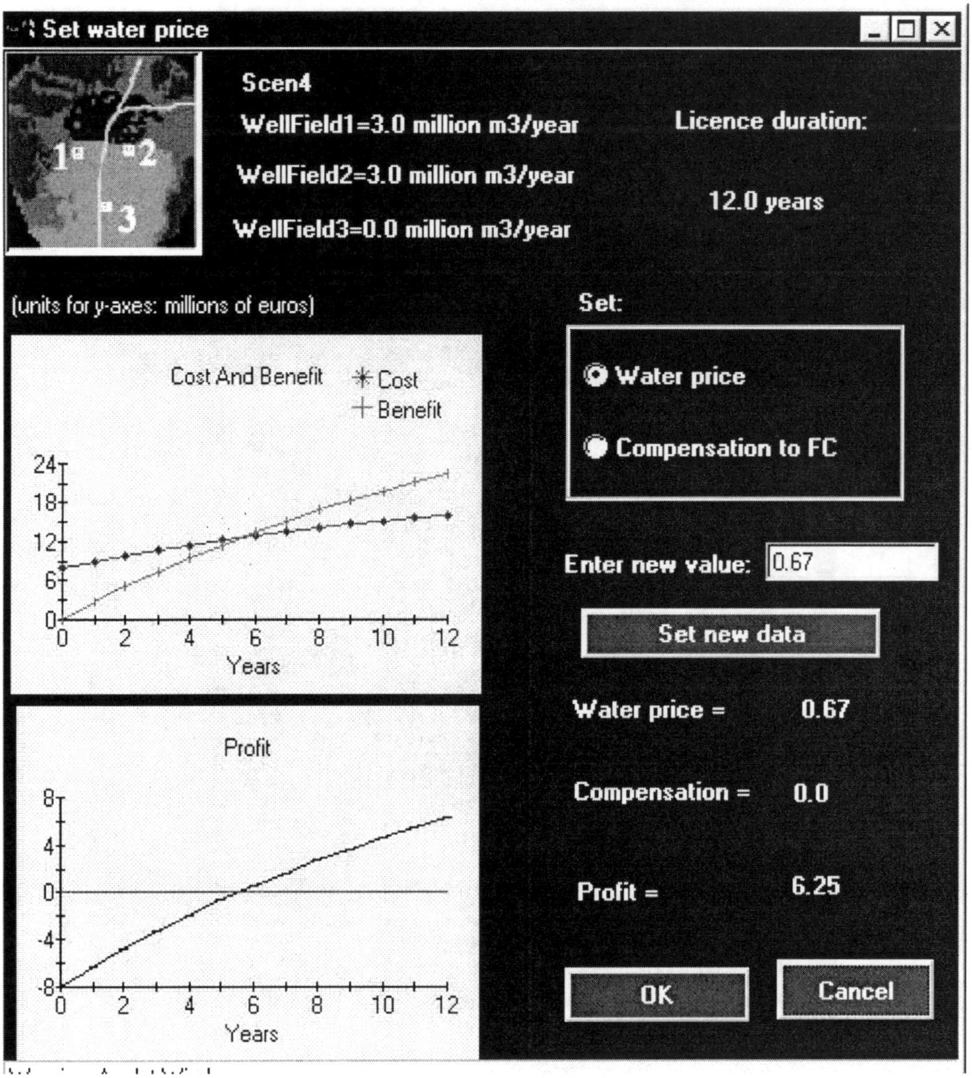

Figure 4.54b. Step 2 (WS) Setting new judgement data by using the economic model

The economic model used contains more parameters then the water price and the compensations for farmers (see again Figure 4.31). By exposing only these two to the role player, the manipulation of data is very much simplified. The same model can certainly be used with different parameters being exposed to the role-player. For example, the discount rate may be exposed to the user so that he or she can test the performance of the scenario with different discount rates. (This may be used for finding the so-called internal rate of return of a given scenario - IRR, which is the rate at which the costs and benefits are equalised). In other cases the coefficient K_{co} may be exposed. This coefficient describes both the real losses of water through the distribution network and the losses resulting from deficiencies in billing and collecting the charges. The influence of these factors on the economic performance of a given scenario can of course be significant.

In any case, once the new data are set for a given scenario, the role player closes the frame and chooses the next scenario from the previous frame. Once all data are set, in a manner similar to the other role-players, they are sent to the server, for storage and eventual updating. The web page for the Water Supply Company role-player is then also created.

Step 3 (Figure 4.55a and 4.55b)

This step is similar to the one of the Local Authority, where the user can check the created web pages for water price and profit (own issues), or for the issues of any other institutional role. Figure 4.55a and 4.55b show how the data in the web page are slightly different for the specific role of the Water Supply Company.

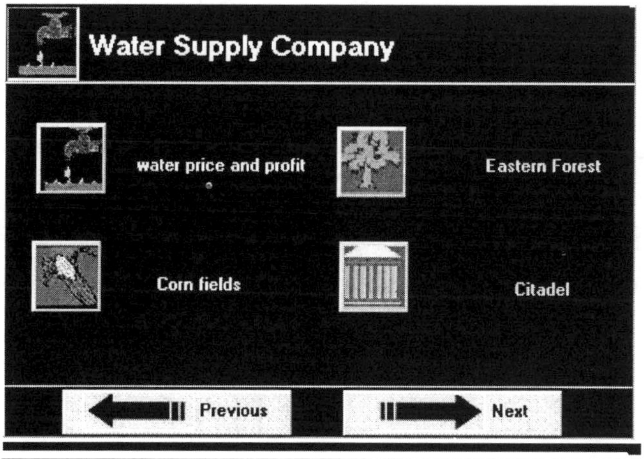

Figure 4.55a Step 3 (WS) Links to the web –pages for the four institutional issues (and role players)

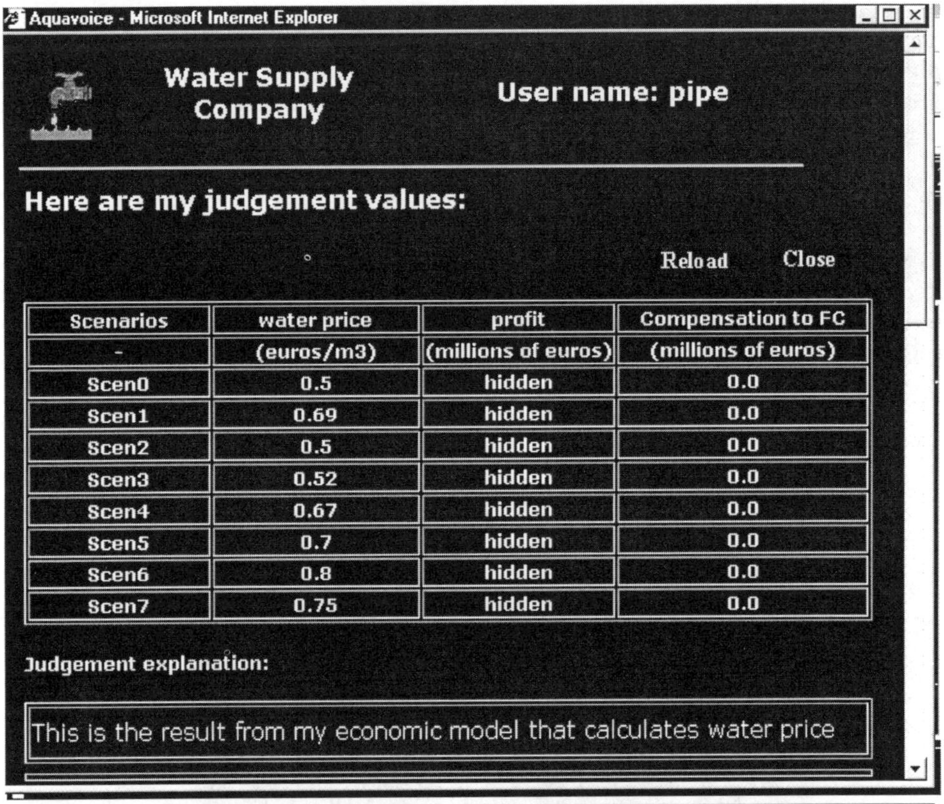
Aquavoice - Microsoft Internet Explorer

Water Supply Company

User name: pipe

Here are my judgement values:

Reload Close

Scenarios	water price	profit	Compensation to FC
-	(euros/m3)	(millions of euros)	(millions of euros)
Scen0	0.5	hidden	0.0
Scen1	0.69	hidden	0.0
Scen2	0.5	hidden	0.0
Scen3	0.52	hidden	0.0
Scen4	0.67	hidden	0.0
Scen5	0.7	hidden	0.0
Scen6	0.8	hidden	0.0
Scen7	0.75	hidden	0.0

Judgement explanation:

This is the result from my economic model that calculates water price

Figure 4.55b Step 3 (WS) Example web page with judgement data of Water Supply Company

Step 4 (Figure 4.56)

In the evaluation stage, this step is almost the same as for the other role players, with the only difference being that this role player has 5 issues instead of 4. The fifth issue is 'profit' and it is probably the most important for this role-player (Figure 4.56).

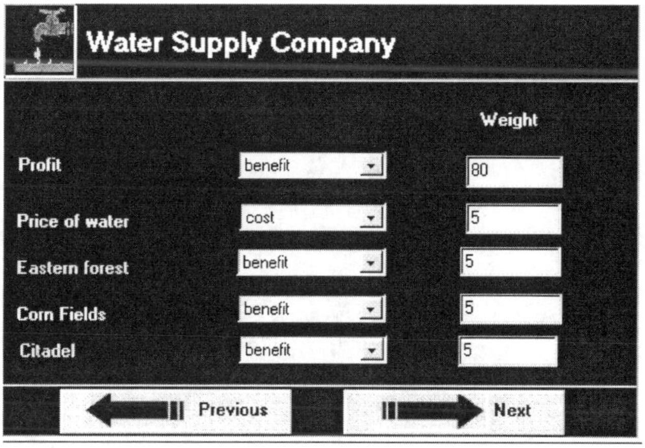

Figure 4.56. Step 4 (WS) Input of evaluation data

The remaining steps of the Water Supply Company role-player are the same as for the Local Authority and will not be presented

Individual User

Step 1 (Figure 4.57)

Any player/participant who joins an *Aquavoice* session with the role of an Individual User has the basic choice of evaluating the scenarios with respect to the four main issues, or raising a new, own, issue (Figure 4.57). If he or she chooses only to evaluate the scenarios against the existing issues, the subsequent steps are same as for the role of Local Authority. The user will be able to see the web pages of the four institutional role players and proceed directly to the evaluation stage. Thus, these steps need not be presented here.

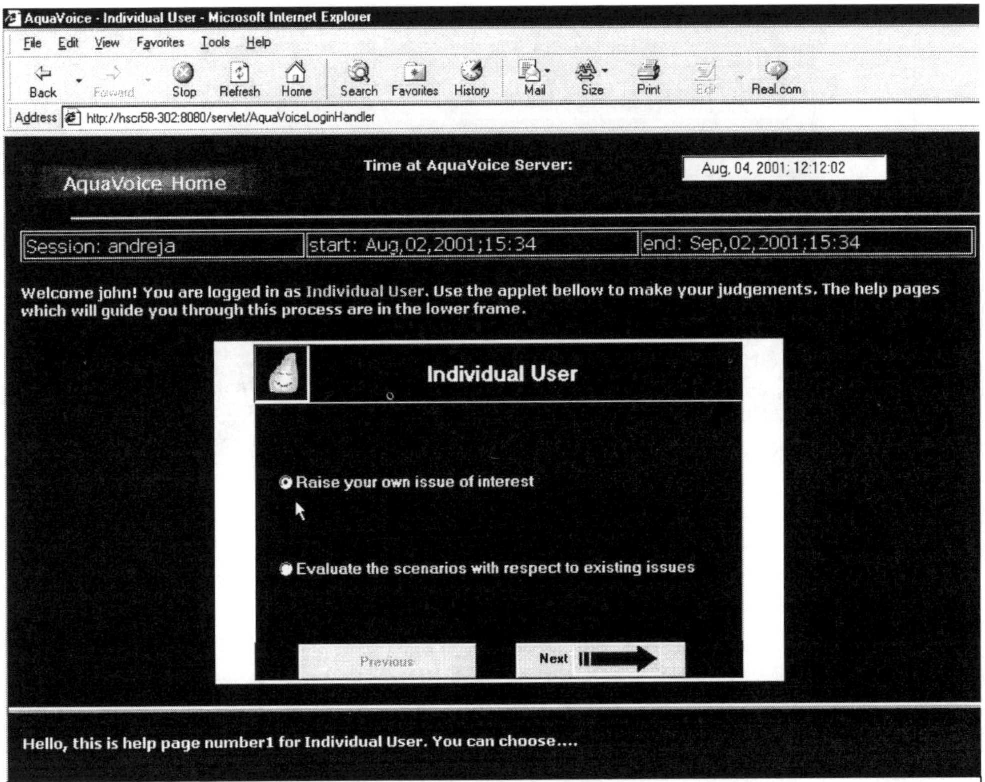

Figure 4.57. Step 1 (IU) choice between raising a new issue or evaluation of the existing issues

In case of raising a new issue, the steps are rather different as compared to the other four, institutional, roles. The subsequent steps that are going to be presented here are only the ones taken when a new issue is raised.

Step 2 (Figure 4.58a and Figure 4.58b)

The new issue that is going to be raised by the Individual User can be one of the listed issues already provided by the system (Figure 4.58a). The idea is that all relevant knowledge related to these issues (in this case the influence of lowered groundwater tables on these agricultural products) can be provided in the help pages that are located in the lower frame of the interface. The Individual User can also choose to raise a new issue, of a kind which has not been anticipated by the system. In that case the first step is to *name* the new issue (Figure 4.58b shows a new issue named 'grass'). It is assumed that the Individual User has his or her local knowledge about the influence of lowered groundwater tables on the new issue, which can later be provided as textual information to the other users.

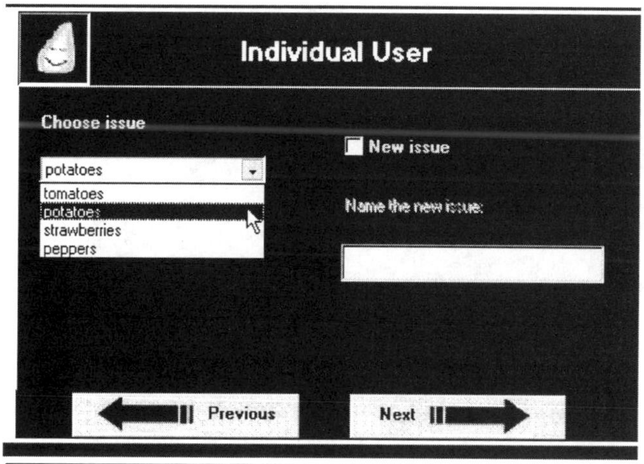

Figure 4.58a. Step2 (IU) Choosing a new issue from the list provided by the system

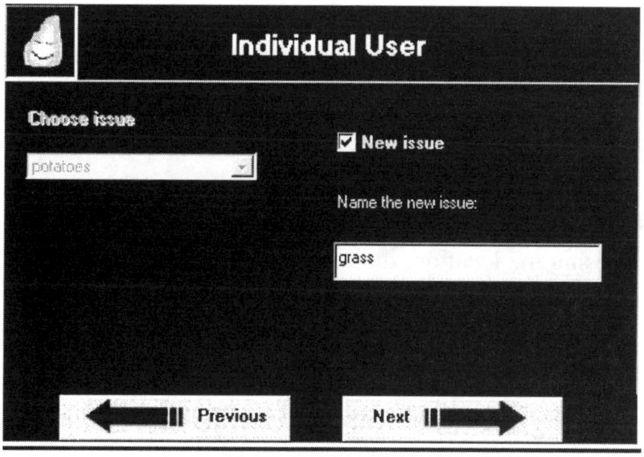

Figure 4.58b Step 2 (IU) Choosing a new issue not anticipated by the system

Step 3 (Figure 4.59)

Similarly to other roles, the Individual user then chooses the kind of MIKE SHE outputs that will be used for judging the impacts on the new issue. Figure 4.59 assumes that the user chooses time series of depth to phreatic surface.

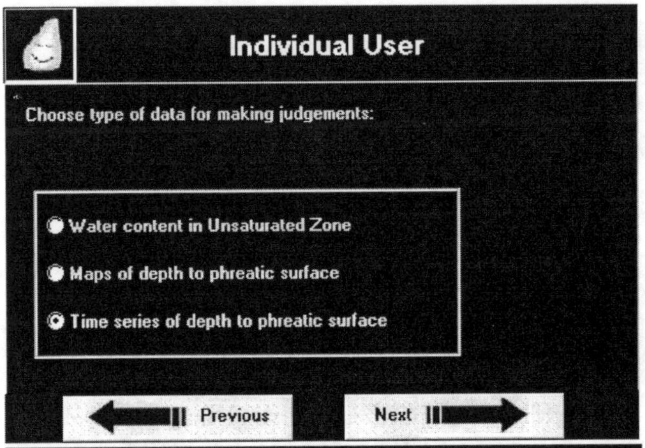

Figure 4.59. Step 3 (IU) Choosing kind of data for judgement of the new issue

Step 4 (Figures 4.60a and 4.60b)

In this step the user needs to indicate the spatial location of the area of his or her interest for which the new issue was raised. This is achieved by using a clickable map of the area (Figures 4.60a). By choosing the location, the user actually chooses one model grid cell for which the MIKE SHE model outputs will be requested from the fact engine. After this, the user chooses the time interval for which the model outputs will be requested (Figure 4.60b).

In case the user chooses to use maps of depth to phreatic surface for making his or her judgements, there is no need to choose the location. In the interface for choosing time, only one date will be active for which the maps will be requested.

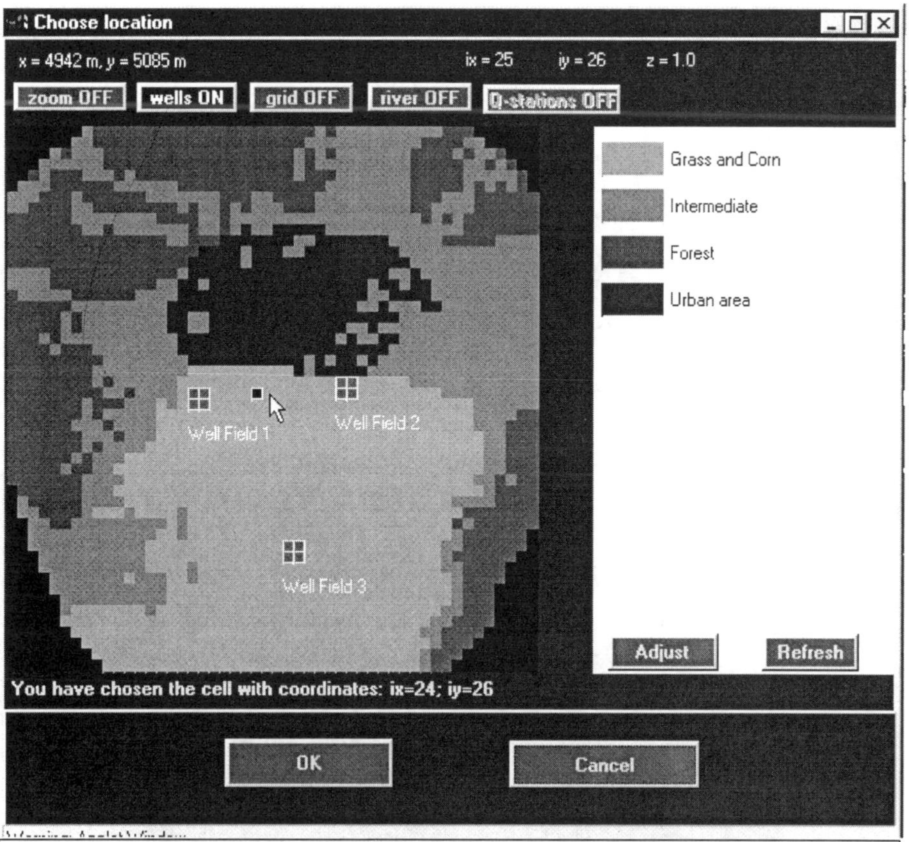

Figure 4.60a Step 4 (IU) Choosing location by using a clickable map

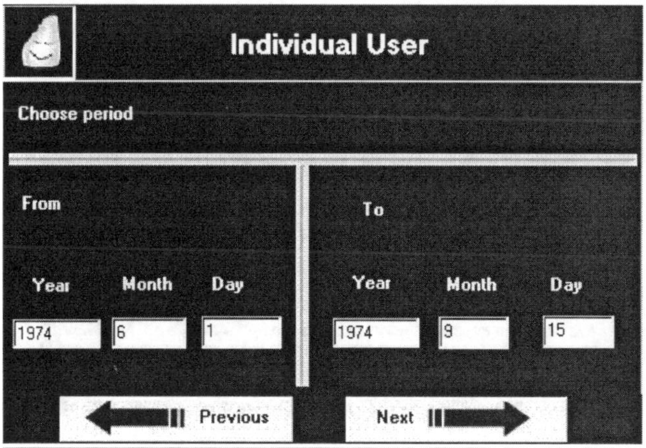

Figure 4.60b. Step 4 (IU) Choosing the time interval for which data will be requested

Step 5 (Figure 4.61)

The role player gets the requested results for all 8 scenarios from the server. In a separate frame any previous judgement data are presented (Figure 4.61).

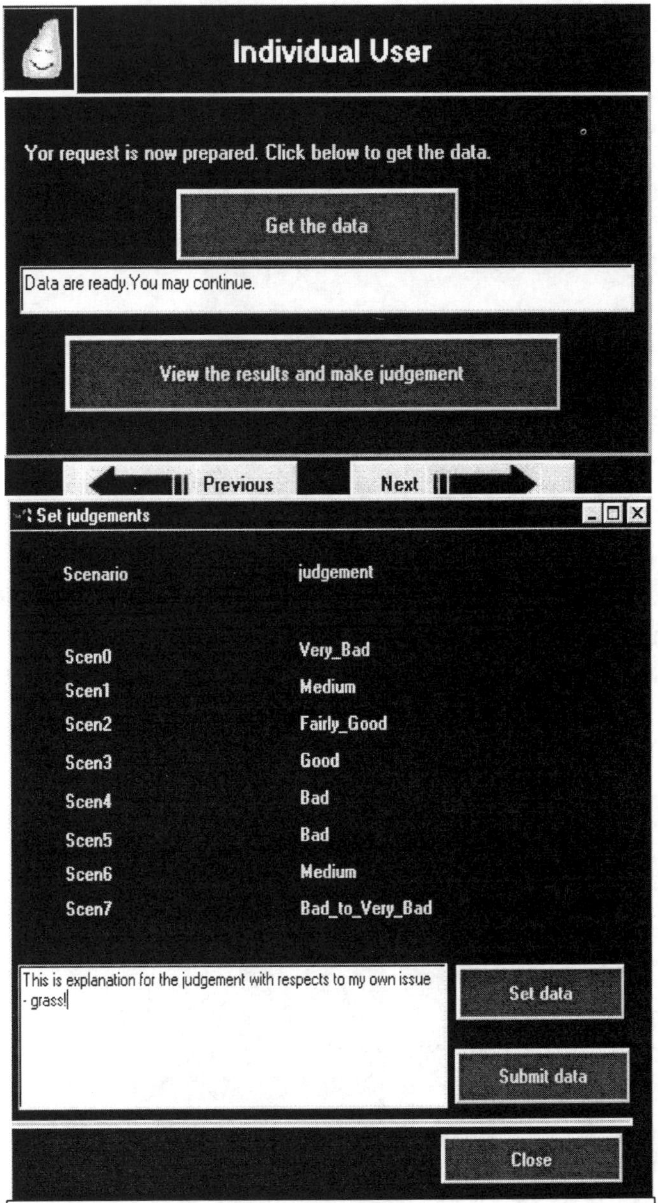

Figure 4.61. Step 5 (IU) Getting the MIKE SHE model outputs and viewing the existing judgement data

This frame also contains the text box for providing the textual explanation of the judgement data. The Individual User may use this text-box to give some extended information about the new issue that was raised.

Step 6 (Figure 4.62)

The user views the data that were requested from the fact engine, and provides the issue judgements in linguistic terms (Figure 4.62). When all new input is provided, the judgement data are sent to the server. The web page for this Individual User is then created or updated.

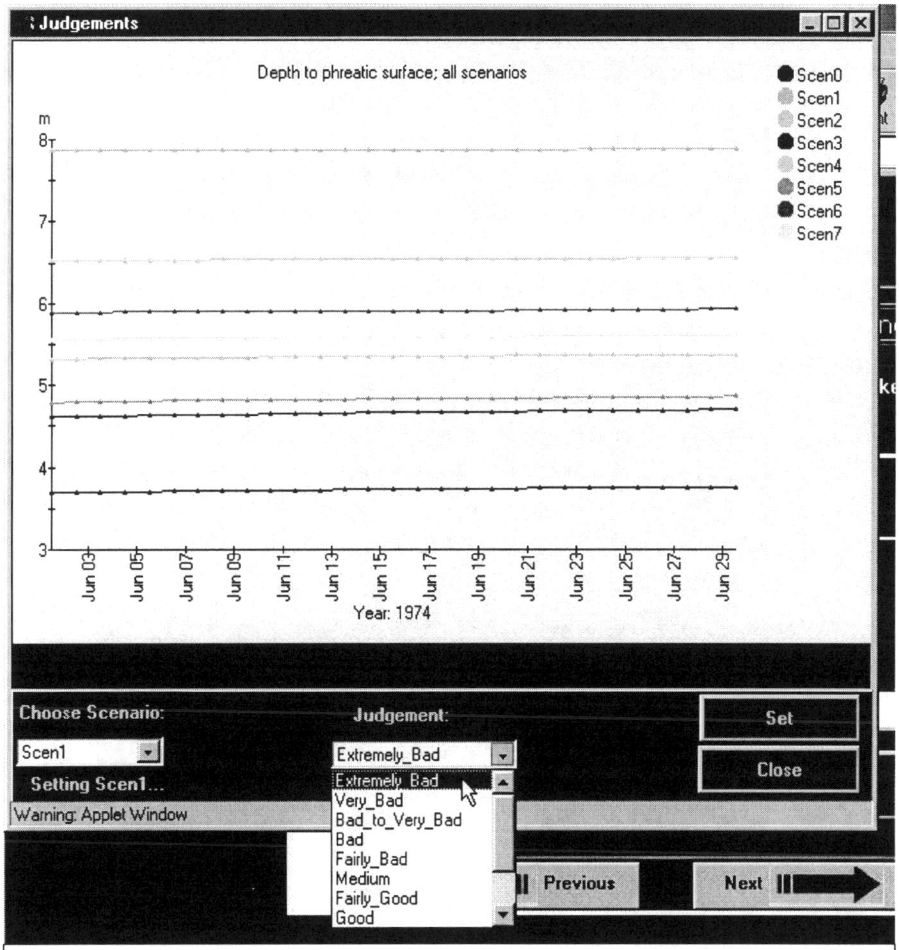

Figure 4.62. Step 6 (IU) Input of new judgement data

Step 7 (Figure 4.63a and 4.63b)

The user accesses the web pages of the 4 institutional role players, and his or her own web page that deals with the newly raised issue (Figures 4.63 a and b).

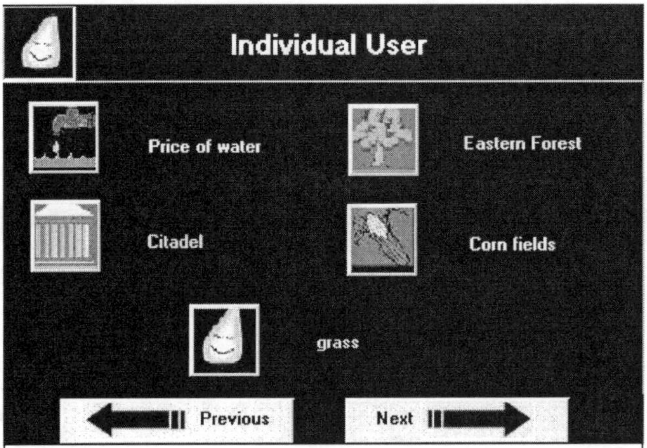

Figure 4.63a. Step 7 (IU) Links to the 5 web pages describing the 4 institutional issues and the 5th, newly raised issue

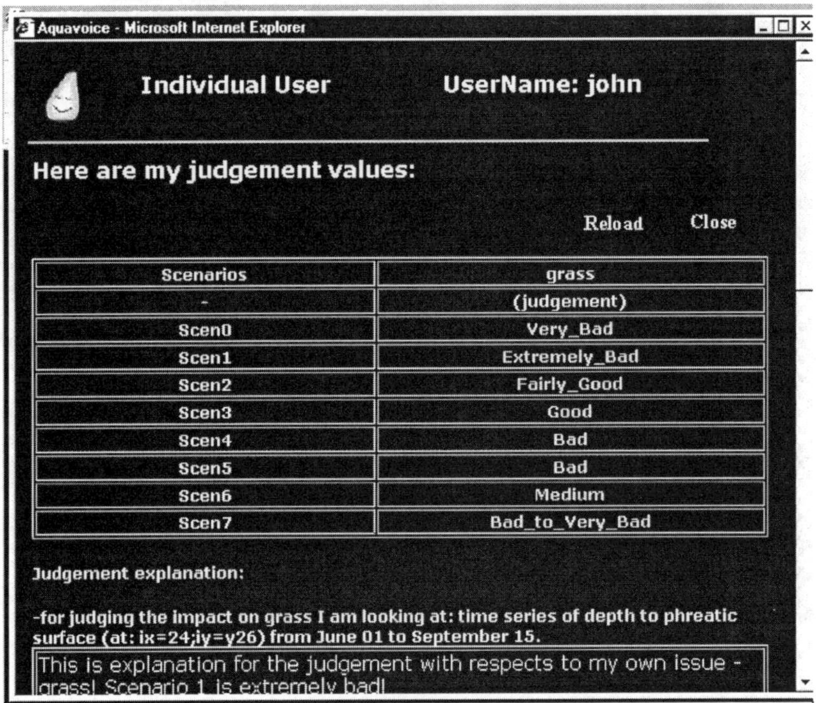

Figure 4.63b. Step 7 (IU) Example web-page for an Individual User

Step 8 (Figure 4.64)

In the evaluation stage the role player with the role of Individual User has 5 issues for which the cost /benefit indicators and the preference weights are assigned (Figure 4.64).

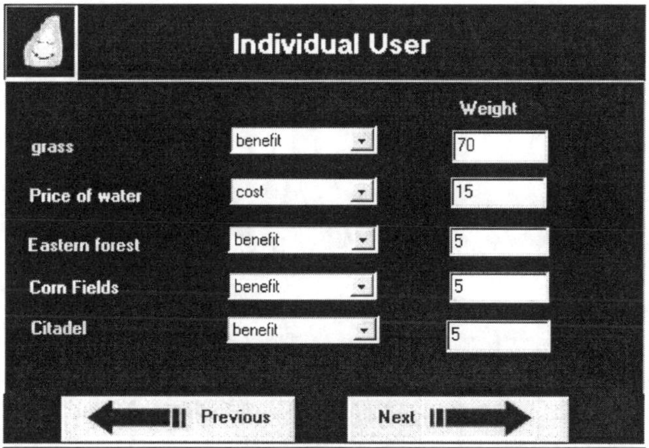

Figure 4.64. Step 8 (IU) Input of the evaluation data

All subsequent steps for this role-player are same as for all the other roles and there is no need to present them. The only interface that will be slightly different for the Individual User is the one where the evaluation updates can be monitored (see Figure 4.65). Here, the top 4 charts represent the ranking of scenarios of the four institutional players, while the chart with the title ‘My Own Ranking’ provides the ranking of the actual Individual User.

The main interface provides a direct access to the web pages of the 4 institutional players. The pages of the players who take the role of the Individual User are accessible only through the chat-like interface for *Aquavoice* discussions. Any Individual User can join the *Aquavoice* discussion in the same way as was described for the role of the Local Authority. Then, the web pages of all players/participants who are currently participating in the discussion can be accessed (by typing ‘url/*username*’; see again Figure 4.65).

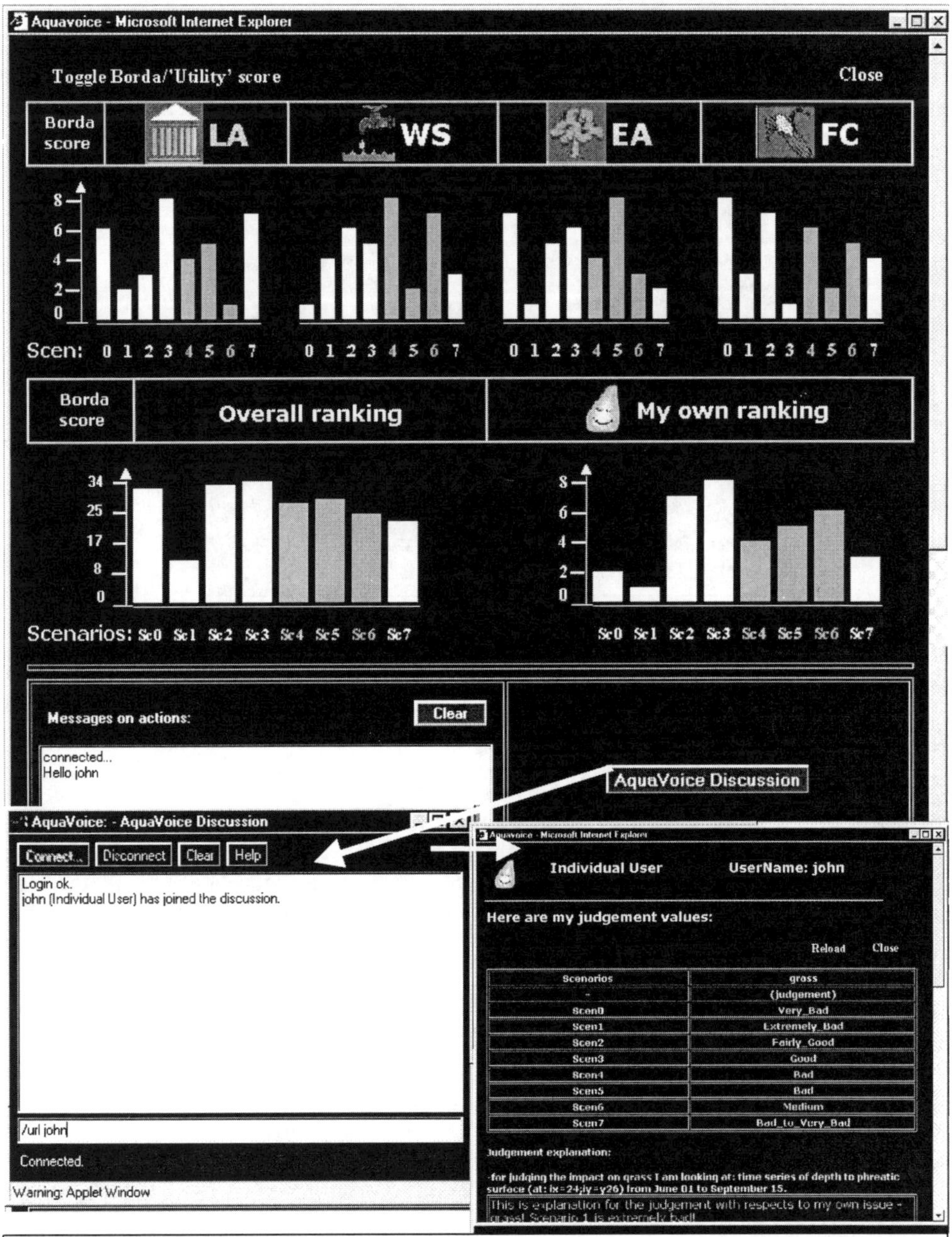

Figure 4.65. Individual User: Watching evaluation updates. Joining the *Aquavoice* discussions allows the viewing of the web pages of all players/participants in the discussion

There is one important point that needs to be raised and which is linked to the steps taken by an Individual User. The option given to this role-player, to raise new issues which were not anticipated during the design phase, seems to open wide possibilities for the abuse of the system. Individual users can in principle raise *any* issue which in *their own view* is influenced by the proposed scenarios of groundwater abstraction. Such an option may certainly be misused by some players/participants in the manner of raising issues related to groundwater abstraction that are complete nonsense (say, to raise the issue of 'colour of my windows' as being affected differently by different abstraction scenarios). Such behaviour can damage the whole cooperative process that the *Aquavoice* system tries to promote.

The question is, then: why is such an option allowed at all to the Individual Users? The system may be designed in such a way that all possible issues that can be raised are pre-determined and simply provided in a list as in Figure 4.58a. Nevertheless, the approach taken here was precisely one of *not* limiting the kinds of issues that an Individual User may raise. Although there are no guarantees that the system will not be abused as described above, we believe that restricting the number and kind of issues that can be raised will be wrong *in principle*.

Firstly, with a limited number of issues which are pre-determined during the design phase, there will be less space for the principal *Aquavoice* goal, which is that of *emergent decision making.* The final decisions will surely *emerge* as a result of the participation of all the players, their concurrent judgements and evaluations, and their subsequent negotiation and collaboration. Yet, such a solution may well exclude a large number of potential players/participants, simply because their issue of concern is not included in the pre-determined list. By allowing the freedom of raising unanticipated issues, many new kinds of groups may evolve, and the possibilities for clustering and alliance building of the potential participants will be practically unbounded. Or, to borrow a term from the developers of computer based role-play games, by letting the Individual Users raise new, unanticipated issues, the control of the *'story line'* of an *Aquavoice* session will be much more in the hands of its users then in the hands of its designer. The features that will support this kind of dynamics are yet to be developed for *Aquavoice*, as will be discussed in the next section, but they should certainly proceed without limits on the number and kinds of issues that can be raised.

Secondly, the very same dilemma is related to the central arguments of this thesis raised in Chapter 2, and is not related only to the development of *Aquavoice*, but to the principle of the approach followed in designing any future NDDSS. If these new network environments aim at overcoming the separation of knowledge and the separation of people, the tendencies for limiting the number and kinds of issues that are *allowed to be raised* must be discouraged. In any real-world decision making problem with potential conflicts that need to be resolved through negotiation and collaboration, *the choice of the issues that are going to be raised is the first, and the most determining, factor*. A system like NDDSS should allow for the autonomous raising of issues of individual or group concern, without any 'outside' limitations. In some sense this is one of the truly postmodern aspects of the design and development of NDDSSs, as introduced in Chapter 2.

4.4 Further Improvements of *Aquavoice*

Through the presentation of the concept, design and development of Aquavoice, we have tried to explain how this system could serve as a prototype of an NDDSS by implementing the concept of a multi-player role-play game.

At this moment, the developed tool serves more as a demonstrator of the basic ideas behind NDDSS development. Its design followed the basic approach and requirements introduced in Chapter 3, subject in turn to the sociotechnical environment outlined in Chapter 2. The three basic components - the fact engine, the judgement engine and the negotiation and collaboration platform - have been successfully integrated in one framework in such a way that the overall system has most of the functionalities set-out in the design requirements. Going back to these requirements we can now see how the solutions implemented in *Aquavoice* meet these requirements.

1. NDDSSs have to enable the access to relevant parts of the body of scientific knowledge, primarily those pertaining to the effects of interventions in the physical domain, but within a given decision-making context. This access should be provided in such a way that various alternative actions relevant to the problem at hand can be tested by any user-participant independently.

This requirement has been met with the fact engine component of Aquavoice. The development and deployment of the MIKE SHE model on the Internet allows for access to the facts related to the decision problem at hand, and for the independent testing of various alternatives by any user of the system.

2. NDDSSs have to enable each participant in the decision making process to express his or her interest and intentions, regardless of whether the participant represents corporate interests, individual interests, or the interests of groups of individuals.

The specific configuration of Aquavoice introduces both corporate (institutional) interests and individual interests. The four main institutional roles are integrated together with any number of Individual Users in one and the same decision making environment. Regardless of their role, all users can have independent interests and intentions.

3. NDDSSs should allow all the participants to combine the relevant scientific facts about the proposed changes in their physical world with their own beliefs and intentions in order to create their own judgement of the proposed and the more desirable courses of action.

The design of the judgement engine component of Aquavoice allows for such combinations of scientific facts and user-beliefs and intentions. The users form their judgements and evaluations of the proposed scenarios by looking at various presentations of the MIKE SHE model outputs. The actual combination of these facts with their beliefs, intentions and interests actually always occurs in the user's mind. The judgement expression is mainly through 'loose' linguistic terms.

4. NDDSSs must provide transparent presentations of all the different interests involved, even if their number is very large.

Particular attention in the design of the system has been paid to the quality of *transparency*. The dynamic creation and updating of web pages with data about all users' positions in terms of their issues, issue judgements, evaluations and textual explanations has been introduced primarily for this purpose. Without such transparency, both the judgement forming process and the subsequent processes of negotiation and collaboration would be practically impossible.

5. NDDSSs need to provide appropriate procedures for aggregating all the different judgements and evaluations, that have been carried out by the participants in the decision making process.

The aggregation of the individual judgements and evaluations in Aquavoice is relatively simple. For the immediate purpose of Aquavoice, the simple summation of the scores from individual scenario rankings is sufficient, and allows for a consistent comparison of each user's individual position with the reaction of the overall community.

6. NDDSSs need to serve as platforms upon which succeeding processes of collaboration and negotiation can be supported, with the ultimate goal of reaching complying decisions.

Aquavoice has a very simple negotiation and collaboration platform. Although it enables these processes, it does not offer any direct support to them. We can say that the Aquavoice design only partly meets this requirement.

Even as this reiteration shows how the Aquavoice design closely followed the NDDSS design requirements, in order to elevate this demonstrator to the level of a fully developed web based system that can be experienced and tested by its potential users it still needs several improvements. Some of these improvements are in fact indispensable for the full deployment of Aquavoice as a multi-player role-play game that can be played over Internet, while others can be considered as subjects for future developments.

A number of these improvements are concerned with functionalities that still need to be developed, while others are purely technical. The first functional issue that is not yet resolved in the system is the means to provide a satisfactory conclusion of an *Aquavoice* session. One of the possible solutions is simply to accept the results from the overall ranking when the time of a given session expires. However, the game could be improved further if it could be extended beyond its time of expiry. In order to have a situation which is more comparable to the real world, at the end of the session, the Local Authority player could make the final decision. This role player might take the overall ranking result as a basis for the decision, but he or she could also decide differently. An interesting extension after the decision is made would be to provide interfaces to all players/participants through which the *behaviour* of the four institutional players could be judged. Any of the four institutional role players can then see how all the others have perceived their behaviour during the session. In real world systems comparable to *Aquavoice,* such features can be used by the institutions involved for an analysis of their own relations with the community.

Another functional improvement would be the extension of the capabilities of *Aquavoice* by running additional scenarios to the proposed eight. As the design of the fact engine has shown, this is in principle possible. The nature of the decision problem may still remain the same, in the sense that it can still be a *choice* problem, but some scenarios that are obviously rejected by all participating parties can gradually be replaced by new, additional, scenarios. Because running the MIKE SHE model over the Internet is time consuming, the additional scenarios cannot be run frequently and by any user. Instead, they should be an outcome of agreements reached during the *Aquavoice* discussion.

The least developed component of *Aquavoice* is certainly the platform for negotiation and collaboration. This component needs a lot of functional improvements. As we have already mentioned, in its present form *Aquavoice* only enables and facilitates the negotiation and collaboration processes and does not yet provide any direct support. The development of a more sophisticated negotiation and collaboration platform requires the implementation of appropriate approaches for measuring 'distances' between users' positions, support for building local alliances and coalitions, and methods for imaginative suggestions of options for mutual gains. These questions are particularly difficult when dealing with large numbers of players / participants and require very much further research. The potential of the newly developed techniques for intelligent data analysis and data mining for knowledge discovery for these purposes needs to be investigated

Among the more purely technical aspects, the issue of security is quite outstanding, as it has not been seriously addressed at all during the development of *Aquavoice*. The simple username / password protection presently employed will certainly not be enough for a system like *Aquavoice* when it is moved into the world of practice. At this moment, the system can be misused and abused in many different ways, and more sophisticated security features will need to be implemented for its protection.

Another technical issue is the development of the help system. Although a framework for using on-line help at any moment during an *Aquavoice* session has been developed, the actual help content is still missing. Without extensive help pages that will accompany all the steps of all the users of the system, *Aquavoice* cannot be properly tested and put into practical operation.

The last technical detail that needs mentioning is the fact that all the back end storage of data about sessions, users and their judgements and evaluations is done using 'flat' ASCII files. A more efficient and more maintainable storage and retrieval would certainly be achieved if the present implementation is replaced by proper database support.

Some of the above mentioned requirements are really needed before the *Aquavoice* prototype can be fully deployed on the Internet and made available for testing. Only after many users have experienced the tool, will other requirements for improvements emerge.

In concluding this Chapter, we need to stress the fact that *Aquavoice* has been developed as a demonstrator of an NDDSS for a specific, imaginary case study, with specific decision problems in mind. Although it satisfies most of the design requirements for an NDDSS, it is not easy to generalise the details of its implementation directly to any other NDDSS. The ways in which new kinds of sociotechnical tools and environments similar to Aquavoice can be constructed and introduced in our societies may still differ a lot from the approach followed here, depending on the actual decision making context. *Aquavoice* should serve, however, as an initial development that introduces the *basic concepts* of the new approach, and as an example of one particular implementation of these basic concepts. With this intention in mind, it can hopefully lead to future, more extensive, projects that will aim at furthering the introduction of NDDSSs into our societies.

Chapter 5

Software Agents and NDDSSs

There are more things in heaven and earth, Horatio,
Than are dreamt of in our philosophy
Hamlet, Shakespeare

5.1 Agents and Agent Orientation

This chapter provides a brief overview of the general field of agent orientation, with the particular technology of software agents as its sub-field, as associated with the potential for applications in the design and development of NDDSSs. Although the emergence of this field has occurred relatively recently, agent oriented studies are by now very extensive, and the concepts are already being explored and exploited in many different areas. In fact, it can be argued that agent orientation represents a significant conceptual shift, to an extent that we can now talk of a new *paradigm of agent orientation*. Because the significance of this new paradigm is now so much broader, we shall begin this chapter with the introduction of some basic notions in agent orientation. After that we shall move to the discussion of the advantages that can be realised with this new kind of approach and which can possibly be used in the design and development of NDDSSs.

Although the study of agents and agency has a long history, the emergence of agent orientation as a new *conceptualisation* paradigm can be attributed to several concurrent developments in different areas that have occurred during the last couple of decades of the twentieth century. Advances in such diverse fields as Distributed Artificial Intelligence (DAI), studies of complex adaptive systems with their applications in different areas such as biology, sociology or economics, network distributed software development, and the explosive growth of the Internet, have all contributed to the development of this new paradigm. The notion of an agent and the understanding of the meaning, the rationale, and the justification for adopting the agent-oriented paradigm therefore vary very much, depending on the respective field of application. Given all the diverse applications that employ various aspects of agent orientation, it is sometimes very difficult to trace their common underlying ground. One possible approach to seeking such a common ground is to look for the origins of this new paradigm in the philosophical, and, in particular, in the epistemological shift that has occurred during the last decades of the twentieth century, which has already been addressed in chapters 2 and 3. It is a shift from a purely *rationalistic* epistemology to a *phenomenological* epistemology.

In the still predominant, rationalistic, epistemology, which is the foundation for the majority of our physical and social sciences, the conceptualisation of any domain of reality proceeds through the identification and *representation* of a number of objects (including concepts) with well-defined properties and relations among them. By systematic experimentation and application of logical reasoning, we are then able to determine the general *rules* that are applicable to given situations in a given domain. The rules are usually coined in (scientific) *theories,* which, once proven, can be applied for predictive purposes. The key characteristic of the rationalistic approach is that the process of cognition of objective reality by the employment of our rational faculties is *independent* of the actual observer, and therefore claims universality.

The second major epistemological school of thought comes from phenomenology. This branch of philosophy, although it does not deny the existence of an objective reality, claims that our cognition of it depends upon our own experiences (experienced as phenomena), which will always depend on our own framework of intentionality (see again section 3.4.2 in Chapter 3). Our cognition of reality does not rely on a manipulation of representations of reality, but occurs through our own experiences in *interaction* with that reality, and in some sense it is an extension of that reality. Using a more technical vocabulary, we can speak of human agents that are *embedded* in their own environment, so that their cognition of that environment is *dependent* upon the interaction of these 'observers' with that environment. The word 'observer', however becomes, itself inappropriate, because here we talk about *intentional agents*, and their interaction with the external environment depends not only on the characteristics of that environment and on their sensory abilities to perceive that environment, but also upon their own, internal, framework of intentionality.

In chapters 2 and 3 we have explored how the limits of the rationalistic viewpoint have already been recognised. The relevance of the phenomenological alternative, although modified along different directions, has already re-emerged in the postmodern schools of philosophical thought. On the other hand, we have already pointed out that the phenomenological method, while having great descriptive power, can hardly be used directly for prescriptive purposes. The introduction of agents and agent orientation can be seen exactly as an attempt to reconcile these two very different viewpoints. In fact it is an approach which tries to accommodate the phenomenological approach within a 'new kind' of rationalistic framework (Devlin, 1995). Although this epistemological background is rarely mentioned in current studies of agents and agent-orientation, in our view it is very important. Both the advantages offered by the new approach and its limitations will be difficult to recognise without an awareness of this background.

What is then the essence of the new approach to conceptualisation that is based on agent-orientation? As a first step, the new approach 'expands' the traditional conceptualisation of knowledge domains beyond their description solely in terms of different objects, with their properties and relations. In fact, it 'replaces' the basic notion of 'knowledge domain' with a much broader notion of an 'environment', where the basic conceptualisation units are not objects, but new kinds of entities called *agents*. The fundamental property of the agents is that they are *embedded* in their environment and they are able to perceive and affect the environment. In addition to that, in their interaction with the environment they express goal-oriented behaviour. This is in fact very close to a general definition of an agent provided by Russel and Norvig (1995) who define an agent as any entity which operates in some environment (physical, virtual, cyberspace), which possesses sensors to perceive this environment, effectors to act in this environment, and goals of its own which may or may not be explicitly represented in the agent itself. In more general terms, an agent can be envisaged as a "creator of objects" within a fluctuating content.

One of the radical changes that this shift in the mode of conceptualisation brings about is that it allows the agents to have *their own* representation of knowledge domains, *which may or may not be part of their own environment*. An agent's environment is therefore not to be confused with a knowledge domain. In some sense, the exclusive role of the (human) observer in the traditional, purely rationalistic, approach, of conceptualising and representing domains in terms of objects, has been now attributed to these new kinds of entities called agents. This is certainly one of the most fundamental changes brought about with the new approach, primarily because it allows for *different perspectives* (or points of view) between the human observer and the agent, and/or between the different agents themselves. In agent orientated studies these different perspectives are sometimes referred to *first person, second person and third person perspectives* (Nehaniv, 1999). The first person perspective is the one of the agent itself, which depends on its sensory and effectory capabilities, and of its internal state (also called the intra-agent perspective). The second person perspective refers to the relations, communications, interactions and the overall experience of other agents, by a given agent (also called inter-agent perspective). The third – person perspective is then the one of the (human) observer, who is engaged in this kind of conceptualisation (also called extra-agent perspective).

Although this epistemological shift can be used as a starting point for the introduction of the paradigm of agent orientation, it needs to be realised that the same paradigm is being developed and applied within several very different disciplines. For example, the very broad area of *computer simulation using agents* has already been applied in many different fields, such as biology, ecology, sociology, economics, business or industrial process management, transport, etc. In all these fields the agent-based simulation is used to observe some *emergent* phenomena, which are resulting from the *individual* actions of large number of individual agents. In nearly all cases the systems that are so modelled are too *complex* to allow the employment of traditional conceptualisation (and the consequent representation and formalisation approach). Of course, in different domain areas, the agents represent different entities (individual organisms in ecological simulations, human agents in social simulations, customers and producers in economic simulations, etc). Depending on the application, the agents and their environments are built with different levels of sophistication. In any case, the central preoccupation of this whole area is the understanding of complex systems and their behaviour at a macroscopic level as a result of the individual actions of a large number of agents acting individually, so to say, at a 'microscopic' level.

This is an exciting area where agent–orientation shows its great potential, and new developments are constantly being reported in previously undiscovered fields of application. The field of water resources management is also one potential application area for this approach. A recent example is the initialisation of a European Project called FIRMA (Freshwater Integrated Resources Management with Agents; for more information see http://www.cpm.mmu.ac.uk/firma/). Our immediate concern here, however, is not with the approach of agent-based simulation. Without devaluing any of the insights that can be gained from this approach, it needs to be realised that agent based simulation has little relation to the design and development of NDDSSs. NDDSS are conceived as environments that can be used *directly* by all interested parties as they participate in the decision making process, and not as *simulators* of the same decision making process: NDDSSs are essentially *socio*technical constructs.

5.2 Software Agents

The advantages of agent orientation for the design and development of NDDSSs are in another wide area, namely that of *software agents*. In the area of software engineering the shift towards agent orientation means primarily the use of software agents as more useful, *higher level mental abstractions,* which can be beneficially used for *software development.* Although many of the issues in software agent technology are similar to the ones that are being raised in agent based simulation, the goals are here very different, namely those of facilitating software development. Software agents need to be seen as a technology that can possibly contribute to the more and more challenging demands put on software applications. Distributed applications, operating in complex and open environments, are typically application areas where software agents can play a role. Building adaptive user interfaces, which can act more autonomously, without total dependence on user actions, are another area where software agents can play a role.

Because software agents can vary so much in their architectures, areas of application and in their respective properties, it is still very difficult to find a common agreement on the definition of a software agent, even among the agent researchers themselves. Different researchers offer different definitions, stressing different aspects of agency as the crucial ones for an entity to be defined as a software agent (compare, for example, Franklin and Graesser, 1996; Nwana, 1996; Wooldridge and Jennings, 1995). The very general definition of Russel and Norvig (1995) which has been stated above, describes well the general notion of an agent, and it may cover everything from humans and animals to robots or simply any programs as agents. Following the approach of many authors who have also experienced difficulties in defining an agent as such, we will here start with introduction of a simple typology of software agents, (which is surely incomplete and may be open to many criticisms). From this typology we shall then try to extract what are the main properties of software agents, and particularly those that can later be used in the design and development of NDDSSs. We shall also see how some design features of software agents directly reflect the inclusion of phenomenological concepts.

Our typology is relatively simple and it classifies the software agents on two distinct bases (Table 5.1). The first column of the table classifies the agents on the basis of the agent architecture, which means *how* are they built.

Table 5.1. A typology of software agents

How agents are built (architecture)	Main functions	Other Properties
Deliberative	Collaborative	Learning
	(Competitive)	Autonomous
Reactive	Interface	Pro-active
	Information	Mobile
Hybrid	(Internet)	Rational

Very broadly, software agents can be classified into three distinct types on the basis of their architecture (Wooldridge and Jennings, 1995). The first group of *deliberative* agents covers those characterised as possessing explicitly represented, symbolic models of the world in which they operate. A deliberative agent makes decisions for its actions through its use of logical reasoning (commonly using first-order predicate logic) based upon the manipulation of the symbols used for the representation of its world. Much of the work on deliberative agent architectures is based on what is known as a classical 'symbolic' AI approach. This approach is faithful to the rationalistic conceptualisation approach, with full emphasis on knowledge representation using symbols. The approach followed in this case for building deliberative agents is comparable to the one used for building expert systems. In fact, this approach to agents also suffers from similar drawbacks, such as brittleness (i.e. an inability to operate outside the narrow knowledge domain for which the agents are built), besides problems with undecidability, and, especially intractability.

These problems have turned many researchers to search for different approaches to building agents. The *reactive* agents are built with no explicit representations of their worlds and no reasoning of the kind that symbolic AI proposes. The reactive agents are built up from a number of low-level modules, which communicate directly with their environment and compete among each other to take control over the agent. This approach originated in the field of robotics (see, for its further expansion in this field, Brooks et al., 1999), but it has been applied in the software agents field as well (Maes, 1994). The reactive agents appear to be much simpler as compared to the deliberative ones, but the complete lack of an internal model in many cases limits their capabilities. A number of architectures called *hybrid* have been proposed where these two approaches are combined in order to overcome the disadvantages of the one or the other.

The second basis for classification of agents in our typology is the main function for which they are developed, or in other words *why* a particular agent is developed. The first group in this classification covers the so-called *collaborative* agents. Probably the most important reason to develop agents is the possibility thereby opened up for cooperation with other agents, in contrast with having just one agent. Systems that are built as societies of agents, where the cooperation and collaboration are exploited for mutual benefit, are in fact known as Multi Agent Systems (MAS). Although multi-agent systems have been mainly developed for collaborative agents, it has to be realised that they may also be *competitive*. In any case it is the *social ability* of the agents that is crucial in the development of MASs.

The social ability of agents in MASs is dependent on the ability of the agents to *communicate* with each other. This is usually achieved by making use of specially designed, highly expressive Agent Communication Languages (ACLs). Agent communication is of such importance in MASs that a number of researchers (Genesereth, 1997; Finin et al., 1997) propose that the ability to communicate through a high-level communication language is the first criterion for agenthood. A whole field of *agent-based software engineering* is now under development which is primarily concerned with the interoperability of different software applications using agent communication languages.

The developments which have probably gone furthest in this direction are the results of the Knowledge Sharing Effort (KSE), a large consortium based upon Stanford University in the USA. This research has resulted in the development of the agent communication language KQML (Knowledge Query and Manipulation Language). This language is based on speech act theory, where, like most of the approaches used in developing ACLs, the message exchanged between the agents is divided into two parts. The first is the so-called illocutionary force (Singh, 1994), which conveys the intent of the sender to the receiver (whether it is a request, question, response etc). In KQLM this illocutionary force is conveyed by the so-called reserved performatives (Finin et al, 1997). They determine the kinds of interactions that are possible with a KQLM-speaking agent. The second part of the message is the message itself, usually called the 'content' of the message that is attached to a performative.

The language is developed together with two other parts of the KSE, which are the language KIF (Knowledge Interchange Format) and the so-called ontologies. The language KIF is envisaged as a common language for all the agents, in which the 'content' of the message is written. Ontologies are the formal specifications of conceptualisation of any particular domain about which the agents communicate (Gruber, 1993). They are the 'vocabularies' that are used to build KIF sentences in a consistent way. Ontologies are in fact very important for inter-agent communication, regardless of the ACL in question, because they contain the descriptions of the domains about which the agents exchange messages.

Although developed together with these two additional features, the KQLM language is designed to work independently as well. For this purpose, each KQLM message carries additional information about the language in which the 'content' is written (which may be an ASCII string, a binary string, SQL, KIF or any other such language), and possibly the ontology that has been employed. In any case, the key feature of KQLM is the separation of the 'content' of the message from the performative (the illocutionary force) which enables communication between KQLM speaking agents regardless of their architecture and internal structure. It can be regarded as a communication layer around any agent which enables its communication and therefore interoperability with other (and possibly entirely different kinds of) agents.

A second major ACL has been developed recently by an agents standards body called the Foundation for Intelligent Physical Agents (FIPA) and is known as FIPA-ACL. This language has much more complex, and therefore more precise, semantics then KQLM. However, KQLM has already emerged as a *de facto* standard ACL, and has been used in many applications. Most tools for building agent-based software are already based on the assumption that the agents will use KQLM for their inter-agent communication. (For an extensive overview of issues related to agent communication, see Dignum and Greaves, 2000)

The second group in our classification covers the so-called *interface* agents. These are not predominantly developed for operation in a society with other agents but as assistants to human users when using a particular application. These agents in some sense radically change the way in which the creation of user interfaces for software applications is understood. Most existing user interfaces are built for tasks that involve closed, static, relatively small and well-structured application domains. Furthermore they rely completely on their user's direct manipulation. Interface agents (sometimes called user assistants, or personal, or expert, assistants), assume that the applications operate in an open, dynamic and sometimes unstructured and unpredictable environment (such as the Internet, for example). Interface agents are active without their users' direct manipulation and they try to match the user's interests, habits and preferences with the changing conditions that they experience in their dynamic environment. Crucial properties of interface agents are their ability to *learn* from the actions of their human users and *adapt* to changes in the environment or to changes in user preferences or behaviour.

The third group in our typology are the so-called *information gathering* agents. These are specifically designed to operate on the Internet on behalf of the user to search and provide the user with a particular type of information. These agents are sometimes seen as extensions of the well-known search engines and are often used for providing personalised or customised content from the Internet. It is obvious that in such applications the information agents can be closely linked with the interface agents. Although they are usually designed to operate as stand-alone applications, they can benefit greatly from cooperation with other agents.

Software agents may be characterised by a number of other properties and some of them may be of considerable importance for applications. The ability to *learn* is crucial for any agent regardless of its architecture or its main function. Learning is in fact a separate topic in agent orientation, and overlaps greatly with the broad sub-field of Artificial Intelligence known as *machine learning*. Depending on the agent architecture, the tasks for which the agent is designed and on the nature of agent's environment, there are different learning approaches.

The learning property is closely related to the property of *autonomy,* which refers to the principle that the agent can continue to operate without direct intervention from the human user or other agents. Without learning capabilities, it is difficult to achieve such autonomy, which is frequently stated as one of the most important conditions for an entity to be considered as a software agent. It is also closest to our common sense understanding of the term agent, as an entity to which we can *delegate* certain tasks and responsibility, which are then carried out autonomously by the agent. In the more general sense introduced above, it is the learning process that leads to the "creation of objects" within an interactive, changing, content.

The property of *pro-activeness* means that the agent will not only react simply to its environment, but it will take initiatives to act in correspondence with its own goals. It is also closely related to the agent's learning ability. In fact without the ability to learn it cannot be said that any agent can show real autonomy and pro-active behaviour. *Mobile* software agents are those which can move through an electronic network from one location to another. This feature in fact introduces a completely different communication paradigm (White, 1997) when compared to agents which are static in themselves but communicate through an agent communication language (ACL). Very often the property of *rationality* is stressed as crucial for a software agent, which means that an agent will always take only actions that lead towards its goals. There are a number of other properties which different authors propose as necessary for a software entity to be defined as an agent. In most cases these properties are closely linked to the particular application of the software agents and because of that they are not usually easily generalised.

As we can see, software agents can be built with different architectures, for different tasks, and can have different properties. When dealing with distributed applications over electronic networks, and especially in open environments, it may be expected that various agents may have to operate under one framework. In such cases we can speak of *heterogeneous* multi-agent systems. The emphasis in the design of such systems is placed heavily on achieving appropriate *interoperability* of these heterogeneous agents. In fact, in many such applications the primary task is the *integration* of various *existing* software applications, known under the name of *legacy software.* The software agent solution to this task is to 'wrap' these existing applications with an 'agent layer', that will ensure consistent and effective interoperability. Regardless of their heterogeneity, the agents will then be able to operate as a society. The key task of this 'agent layer' is certainly to provide all the agents with common communicational capabilities. The ACLs discussed above are designed exactly for this purpose.

In addition to the capability to use ACL, these agents, although still heterogeneous, can significantly improve their interoperability, and their general operation as a society, if they posses the capability to reason with their own knowledge and with the knowledge of the other agents. Yet, at the same time, they have also to be able to preserve their autonomy and control over their own actions. The general design approach usually taken for such agents follows the so-called 'strong' approach to agency which is favoured by a large part of the AI community and is connected with the issue of the 'mental state' of an agent.

This strong notion of agency takes the stance that artificial agents in general (and in this particular case, these are software agents) should not only be conceptualised, but also implemented using concepts that are more usually applied to humans (Wooldridge and Jennings, 1995). For this notion to be viable, it is quite common to characterise an agent using 'mentalistic' notions such as knowledge, belief, intention, desire, obligation etc. This stance is known as the *intentionalist stance*, a concept taken directly from phenomenology. This approach is in fact a direct example of bringing phenomenological concepts into a new kind of rationalistic framework.

Although we normally use these mentalistic notions to explain and predict human behaviour (e.g. "John went to see his boss because he *believed* he would listen to him" or "Mary studied hard because she *wanted* higher marks"), it is not uncommon that we refer to many artifacts as well using this same notion. We say that the cash dispenser *knows* the balance on my account and it does not *want* to give me money. Or that the computer *believed* I wanted to print the file. Taking this approach further in using these mentalistic notions (usually called mental *attitudes)* for conceptualisation and implementation of artificial agents may seem too anthropomorphic, but it has some clear advantages.

The usefulness of this approach has been indicated already for stand-alone applications (McCarthy, 1979//1990), but in the case of multi-agent settings mental attitudes can actually be used as high-level design tools. Firstly, as we have already said, the multi-agent systems may consist of heterogeneous agents, designed and programmed by different people. Then the notions of mental state may be very useful in order to achieve some coherent interaction between the various agents. An agent may, for example, observe the behaviour of other agents in the same environment and induce information about their plans, goals, beliefs or intentions. In the case when the agents communicate via ACL, they can communicate this kind of information directly to each other. In that case it is of particular importance that all the agents which are in the framework of interaction be built of 'material' like beliefs, goals, intentions etc. An agent which is not built in this way will have problems to articulate any of these attitudes (Shoham, 1994). Although there are many applications where agents communicate through an ACL but are constructed without a mental state, many researchers point out that the use of ACL is, in fact, inseparable from the mental state of an agent. Communication using ACL consists of *communicative acts* which, like any other actions, are *intentional*, which means that they are dependent on the mental state of the agent. The notion of mental state is most radically accepted in the paradigm of Agent Orientated Programming proposed by Shoham (1997), where the state of the agent consists entirely of mental attitudes, such as beliefs and obligations.

There is no agreement, however, on what is the proper choice of mental attitudes to build the mental model of an agent. The so-called "BDI-architecture" (Beliefs, Desires and Intentions) only opened the way to the realisation that a combination of mental attitudes can lead to entirely new and different mental models. Indeed the choice of the attitudes for building the mental model of an agent is usually driven by the particular application for which the multi-agent system is developed. However this approach is at the same time associated with a large set of problems that still have to be resolved. We can just stress here that using mental attitudes in some logical frameworks inevitably leads to extensions of classical first-order predicate logic with other logics that are usually called non-standard logics (modal logic, temporal logic, deontic logic, etc.). These extensions introduce new kinds of logical operators which should capture the common-sense understanding of various mental attitudes, which are not truth-functional in the manner of the operators of a standard first-order logic. This causes immense problems and, although many different approaches have been proposed for handling mental attitudes, even the number of problems is still far from settled.

At the and of this introduction we need to say that software agents are still not widely accepted as an alternative for building distributed software applications in open environments. There are several reasons for this situation, and one which cannot be ignored is the so-called *overselling* of the technology with promises that it may become some kind of a 'silver bullet' that can resolve all kinds of software development problems. There are still some fundamental problems that cannot be easily resolved in software agent technology, and they are already being recognised by many software agents researchers. An excellent overview of the developments, and the expectations from this technology, as contrasted with its current reality, is presented in Nwana and Ndumu, 1999. Among the other issues that they raise, probably the most significant is the so-called 'ontology problem'. It relates to the fact that the creation of an ontology for a given domain, which can then be used by various agents, still requires an *explicit definition of every concept that needs to be represented.* General-purpose ontologies are therefore still practically impossible, and this turns the developers towards building domain-specific ontologies. There are currently many efforts in this direction, but it needs to be realised that agent applications are not only built for different domains, but also for different *tasks*. This is what Nwana and Ndumu (1999) refer to as the *interaction problem*. Same domains for different tasks require possibly different ontologies. This in fact touches upon some fundamental problems of knowledge representation. In addition to this, different domain-specific ontologies will require converters to translate the concepts from one ontology into another, which may further complicate the agent communication and cooperation functions.

Apart from the unresolved theoretical issues, another frequently reported reason for the slow acceptance of software agent technology by the software engineering community is the fact that there are still not many problems that require such sophisticated approaches for their solution. For many software development problems, the capabilities of the more powerful object-oriented languages are enough. Their extension with various distributed objects technologies has in fact made them powerful enough to tackle most problems, even for distributed applications over electronic networks. When it comes to the design and development of NDDSSs, we can surely expect that, for different kinds of decision making problems and environments, different approaches will be used. However, the way in which these systems have been introduced here suggests that because of their complexity, openness, and necessarily distributed nature, their design and development is a task for which the software agents approach can possibly be very suitable. In what follows we shall then try to link the design and development of NDDSSs with software agent technology.

5.3 Possible uses of software agents for designing NDDSSs

Let us first try to provide some general justification for the proposal that software agents are useful for building NDDSSs. We shall follow the general recommendations from software agent researchers concerning the key characteristics of application domains where this approach should be considered and then see if they correspond to the domain of NDDSS construction. It is commonly agreed among the software agents community (see, for example, Jennings and Wooldrige, 1998; Weiss, 1999) that this approach is suited for applications that are *distributed, open* and *complex*.

Distributed applications are those which require a number of different software components (programs, data sources, control components etc), which can be physically or logically (conceptually) distributed. In such applications, it is usually the case that, in order to solve problems, these components need to interact with each other. In these situations the software agents approach can provide a suitable alternative. At the high abstraction level of the agents it is irrelevant whether the distribution is conceptual or physical.

NDDSSs are distributed systems both in the physical and in the conceptual sense. They are physically distributed over electronic networks (like the Internet), and they consist of functional components that are conceptually distributed. For example, one physical location (one machine) may host several different kinds of fact engines in the knowledge centre, each responsible for the provision of different facts (for example, outputs of several different kinds of models). On the other hand, one generic type of judgement engine component may be distributed over a number of different client computers, and serve many users. Regardless of this different *type* of distribution of all components in an NDDSS, the design should guarantee their proper interaction. Conceptualising the different components as software agents elevates the analysis above these lower level considerations and can be beneficially exploited in the design phase of an NDDSS.

Open application domains are those where the overall structure or the number and the type of the required components may change over time. Usually such application domains are characterised by a very great heterogeneity of the components involved. (for example, programs that are developed by different people and organisations, at different times and with different languages). This situation is obviously closely linked to problems of integration of legacy software into a framework where proper interoperability can be achieved. The best example of an open software application environment is the Internet itself, with a huge number of different software and information resources developed by different people and organisations. Exploiting such resources is the most challenging demand that is presented to new software applications, and involves issues like communication, cooperation and negotiation between independent software components. Conceptualisation and/or implementation of these components within a multi-agent framework can be especially well suited for these situations. However, it should be noted that it is exactly the openness of the application domain which underlies the 'ontology problem' discussed above.

Although NDDSSs will be initially designed and developed for particular decision making contexts, we have still to identify them as open systems (see again chapter 3). We have defined NDDSSs as knowledge circulation environments, which cannot be seen simply as "tools to be applied, but processes to be developed". As the number of their potential users is envisaged to grow and become very large, the system will have to be able to meet new and increasing number of interests and requirements. In fact, this means that, up to a certain level, all three functional components of a given NDDSS will have to be designed as open. Even if the way in which this openness is approached may be different in the three different functional components, the system as a whole will need to be able to cope with this openness. As we shall see later, probably, the best approach for dealing with it is to employ different kinds of agents within different functional components of the system.

Complex application domains are those where a large number of components are operating concurrently and in an asynchronous and non-linear manner. The best way to deal with such domains is to employ a *modular* structure. The domain is then first decomposed into specialised components, each of which deals with particular aspects of the domain and thus is easier to develop and maintain. The issues related to the interactions and interdependency of these independent modules are then addressed as cooperation problems. Agent orientation is especially suited for these kinds of situations because each software agent can use the most appropriate approach for solving its particular task, independently of other agents, while the whole society of agents can use a common communication and cooperation framework for dealing with the overall complexity.

From what has been presented in previous chapters it is obvious that the complexity of NDDSSs necessitates such a modular approach. These systems need to support concurrent, asynchronous and non-linear knowledge circulation processes. As we shall see, in different functional components we can identify tasks which can be handled by different classes of agents, while, at the same time, some of them will have to have many instances in order to serve the large number of potential users.

Let us now see more specifically what are the kinds of agents that can be used in the three functional components of an NDDSS, and what are the approaches available for their integration into one multi-agent framework within and across these components.

5.3.1 Agents in the fact engine component

As already indicated in section 3.4.1 of chapter 3, one of the primary problems that will have to be resolved in the fact engine component of a given NDDSS, and for which the agent oriented approach may provide a suitable approach, is the problem of software interoperability. There are two important areas related to the functionality of the fact engine component where software interoperability arises as primary concern. The first one is the already stated demand for *integrated modelling*, especially in view of the wide range of knowledge needs that the system as a whole will have to satisfy. Many, if not most of the fact engines that can provide valuable facts to the many users of an NDDSS are envisaged to be various kinds of models. Most of the existing models are separate models, designed to deal with particular classes of problems. The example of MIKE SHE, which was used for the development of the Aquavoice fact engine, is still rather unique, as it integrates all the land based- hydrological processes into one computational framework, albeit at the cost of maintaining the interoperability of the components within its own structure. Obviously, such a model greatly reduces the problem for software interoperability. Because of the increasing demand for integrative approaches in water- and environment related decision making in general, there is already a pronounced orientation towards the integration of separate models, which will continue irrespective of NDDSSs development. It has to be realised, however, that the potential number of models, and the level of their integration that might be required for a fully developed NDDSS, can be very large.

The second area where the fact engine component encounters software interoperability problems is the potential existence of various software components that will indeed be very heterogeneous. Databases, knowledge bases, documents, expert systems, and any number of other information sources or software applications, sometimes possibly not related directly to the fact providing tasks, but still supporting them, will need to be integrated within one framework.

For both these areas, the multi-agent approach for achieving software interoperability seems advantageous. In particular, the use of autonomous agents equipped with communication capabilities using ACL can be employed. Existing software applications may be converted into agents, by 'wrapping' them with such a communicational layer, while new applications can possibly be equipped with communicational capabilities of this kind from the beginning.

Such a design approach for building the fact engine components of an NDDSS would mean that each component would then be converted into an agent, basically capable of two kinds of actions, which we can now call *private* and *communicative*. The private actions are those related to the internal tasks for which a given software component is built (execution of a model run, extraction of a query result from a database or a knowledge base etc.) For its private actions each agent operates within its own, internal, and rather restricted domain, and the way in which such agents are implemented is independent of the particular layer in the agent communication process.

At the same time, each agent is capable of sending and receiving messages in a common ACL, such as KQML. In order to be able to exchange meaningful messages, the agents need to use a common ontology (or several common ontologies). It can be envisaged that the fact engine component, which is located in the knowledge centre of an NDDSS, will still operate in a relatively controlled and bounded environment, where both the domain and the tasks can be well defined, and the use of common ontologies will be feasible.

In addition to the capabilities that they mobilise to execute private and communicative actions, the agents need to achieve a proper level of coordination of their actions. The ways in which this can be achieved can vary, depending on the actual tasks at hand and the implementation architectures of the agents involved. A number of different approaches have been proposed for achieving proper agent coordination and cooperation, such as the use of private and social commitments and conventions, cooperation and negotiation protocols (such as the contract net protocol or the blackboard systems), or implementation of formalised negotiation procedures. Many researchers suggest that only agents that are implemented using the 'mental state ' approach can make an adequately sophisticated use of both ACL and/or such cooperation and coordination protocols. In the case of agents employed as components of the fact engine component of a given NDDSS, the decision about the choice for the actual implementation architecture of the agents of the fact engine component will depend upon the requirements of the particular application.

5.3.2 Agents in the judgement engine component

When it comes to the judgement engine component of an NDDSS, the problems are of a completely different nature. The primary requirement of the judgement engine component is to help the users of an NDDSS to form their judgements on the basis of the relevant facts received from the fact engine component, and the *users own* beliefs, intentions and interests. As we have already mentioned, this requirement places the emphasis of the judgement engine design upon their user interfaces. The actual choice and combination of signs and sign vehicles that will provide an appropriate presentation of the required facts, and will allow the users to form their judgements, will be dependant on the particular application. Dealing with these design aspects seems to have little or no relation to the use of software agents. It has to do much more with agents in the sociotechnical sense of object-creating entities that are combinations of human and artificial 'materials'.

However, the way in which the judgement engine component is supposed to function within an NDDSS as a whole, points to other aspects of its design where the use of software agents can still be of value. Firstly, in order to achieve its main function, the judgement engine component has to provide customised and even personalised interfaces for users who can have very *diverse* beliefs, intentions and interests. The obvious candidates for these tasks are the so-called interface-agents (or 'user assistants'). At the same time, the whole 'background' against which the users are forming their judgement is dynamic and may change over time (users with new kinds of interests join in, new facts are being provided, the social landscape changes as a result of negotiations, etc). This means that these interface agents will have to deal both with the changing *attitudes* of their users, *and* with such changing situations in this 'background environment'. These aspects make the judgement component of an NDDSS a truly 'open' application domain, being open also to its human agents.

Building interface agents capable of dealing with such situations is very difficult. The most recommendable approach would certainly be if these agents could be built with the abilities to *learn* and *adapt* to continuous changes in both the user attitudes and in the 'background environment'. The advantages from having learning and adaptive interface agents will be twofold. Firstly, such interface agents will be able to act as true 'personal assistants', as they will be able to suggest beneficial actions to the user in a changing dynamic environment. Secondly, such agents would in principle be capable 'replacing' the human users and continuing to act *autonomously* on their behalf when they are absent. The achievement of this second advantage involves the difficult issue of *trust* that the human users can put on such autonomous applications. A lot of experiments and prototyping are required with multi-player games like *Aquavoice*, extended to include such autonomy of the agents, before they can be included in systems supporting real world decision-making problems. Once again, we have to do here with agenthood in the sociotechnical sense rather than in the purely technical sense in which it is usually conceived in the software engineering and computer-scientific literature.

It has to be realised, however that building interface agents with such sophisticated learning and adaptation capabilities is extremely difficult. As Nwana and Ndumu (1999) point out, the design of such agents has to meet the so-called 'intelligent tutoring and mentoring problem', for which the agent is required to have four key knowledge modules: domain knowledge on the application, a model of its user, strategies for assistance, and some sort of a catalogue of the typical problems (situations) that arise within the particular environment. These requirements suggest that such agents need necessarily to use deliberative architectures, which however are very difficult to built. The alternative approach of Maes (1994) of using more reactive architectures built out of competing competence modules which produce emergent learning and adaptive behaviour, have been successful in several applications, but in domains that are still rather simple. This area is still under active research, but it seems that some kind of hybrid architecture will be needed in order to meet the requirements of complex application domains like the judgement engine component of an NDDSS.

Because such applications are still an intention, rather than a reality, the judgement engine interface agents can, as in many other cases besides, take a more simplified approach. Instead of continuously learning every 'move' of its user, the agent can use the so-called 'user profiles' created from the user input, and than try to 'match' situations occurring in the 'background environment' with such representations of the user. This approach is similar to applications for providing customised content, as exemplified by many existing Internet applications.

Regardless of the approach taken for building the interface agents, one outstanding problem is the way in which they will be connected to the heterogeneous agents which constitute the fact engine component of a given NDDSS. With the envisaged large number of interface agents that will serve all the users/participants, and the diversity of the facts that they will require for supporting the judgement forming process, it may become very difficult to maintain *direct* connections between the interface agents and the fact engine agents. Such direct connections may be very expensive in terms of computational resources, and thus increase the complexity at the implementation level prohibitively. This is the already well-recognised problem of *scalability* of distributed multi-agent applications.

In order to overcome problems of this kind, one common solution is that of employing another kind of agents, composed of so-called *facilitators*. Instead of using direct communications, the interface agents communicate through these agents, which thus act as their intermediaries. The facilitators, envisaged to be located in the knowledge centre, establish and maintain connections across all interface agents and ensure a coordinated exchange of messages between them and the heterogeneous agents acting as fact engines. In communicating with the fact engines, the facilitators then need to be capable of using the same ACL as all other heterogeneous agents. All the agents involved can then be organised in a so-called *federated system*, where the communication overhead is significantly reduced. As we shall see, the use of such a federated system using facilitators is also advantageous for the collaboration and negotiation platform.

5.3.3 Agents in the platform for collaboration and negotiation

In chapter 3 we have already pointed out that the collaboration and negotiation processes within an NDDSS are essentially carried out by the *human* participants. In such complex situations involving 'many parties and many issues', the complete automation of these processes is extremely difficult, if not practically impossible. The purpose of the platform for negotiation and collaboration as a component of an NDDSS is then primarily to *facilitate* these processes. It has already been repeatedly stressed that an NDDSS is a sociotechnical system comprising human users and the software system, the two held together in an *interactive* framework. While in the judgement engine component the focus is mostly on the interaction of the *individual* user with the software components, the platform for negotiation and collaboration has to be designed for the interaction of the participants as a *group*. More precisely, this component has to support the expansion of the judgement formation process from the domain of the individual to the domain of the group.

In order to provide such support, the negotiation and collaboration platform needs to have components both at the users' periphery and in the knowledge centre. At the users' periphery, the software requirements are again primarily related to the design of user interfaces through which the participants can join the negotiation and collaboration processes, while monitoring their progress. This means that the *interface* agents, already introduced as a possible approach for constructing the distributed judgement engine component, can have their role extended so as to provide user assistants in the collaboration and negotiation processes. However, because the emphasis here is on supporting the individual user with respect to his or her position in the group as a whole, the interface agents need to be extended so as to provide both the social landscape emerging from the interactions of all participants and the embedded position of that particular user in such a landscape. For this purpose, the interface agents need to act as a kind of 'medium' through which the 'moves' of all the participants are made transparent. This transparency should enable the users to discern the actions by which they can orientate their attitudes relative to one another, and therefore enhance the collaboration process. This implies that the interface that is actually presented to each user will depend on the actions of all the other users/participants.

The actual signification approach employed will again depend on the particular application at hand. It needs to be stressed; however, that regardless of the interface agents representing the same social landscape, and supporting the same kind of tasks for all the users, the choice of sign vehicles employed may be different for different users. The issues of customisation and personalisation therefore remain just as important as when we discussed the interface agents for the judgement engine, although here the emphasis is more upon the *form* of the representation, rather than upon the *content*.

Because the representations in the negotiation and collaboration interfaces for each user of a given NDDSS will eventually depend on actions of *all* users/participants, the interface agents need to be interconnected in such a way that they can communicate with each other. Here the issue of scalability, related to direct communication between the interface agents, arises again, and now in an even more severe form, as this implies that all agents will need to employ a so-called 'broadcast message-passing' approach. The solution needs to be sought again in the federated system, that has already been proposed, with the use of facilitator agents that act as intermediaries between the interface agents.

Apart from this basic technical task of coordinating the communication exchange between the interface agents, however, the facilitator agents of a collaboration and negotiation platform of a given NDDSS need to encapsulate all the necessary knowledge for supporting the collaboration and negotiation processes. These agents therefore need to take upon themselves the necessary tasks of aggregation of the individual judgements and positions, as well as the creation and updating of the social landscape presented to the users through the interface agents. The support in terms of calculating *similarities* of users' positions that can be used for coalition and alliance building, or the determination of alternatives with mutual gains, are also tasks to be handled by the facilitator agents. The facilitator agents may integrate these additional tasks within their own agent architecture, or they can maintain separate connections with other 'back-end' agents specially built for this purpose. With these tasks in view, it seems that the facilitator agents, although envisaged to be relatively few in numbers, may actually provide an essential support for the negotiation and collaboration processes.

An overview of design objectives of each NDDSS component, their specific tasks, and the types of agents which can support them is given in table 5.2

Table 5.2

NDDSS component	General objectives as specified by design requirements of an NDDSS	Specific tasks to be supported	Type of agents
Fact engine	Fact provision, access to relevant scientific knowledge	Integrated modelling; software interoperability; data processing and assimilation	Heterogeneous agents communicating by using common ontology and common ACL, possibly with mental states
Judgement engine	Enabling individual judgement formation, with respect to available facts, and with respect to judgements of other users/participants	Customisation and personalisation of both form and content; Transparent presentation of individual positions within the 'social landscape'	Interface agents acting as personal user assistants, equipped with learning and adaptive capabilities
Platform for negotiation and collaboration	Facilitation of the negotiation and collaboration process	Coordination of communication exchange; Support for coalition and alliance building	Facilitator agents acting as intermediaries; Back end agents for calculating 'similarities' of users' positions

As a summary of this discussion about the potential use of software agents for building NDDSSs, we can present one possible, conceptual, agent-based architecture of an NDDSS, as shown in Figure 5.1.

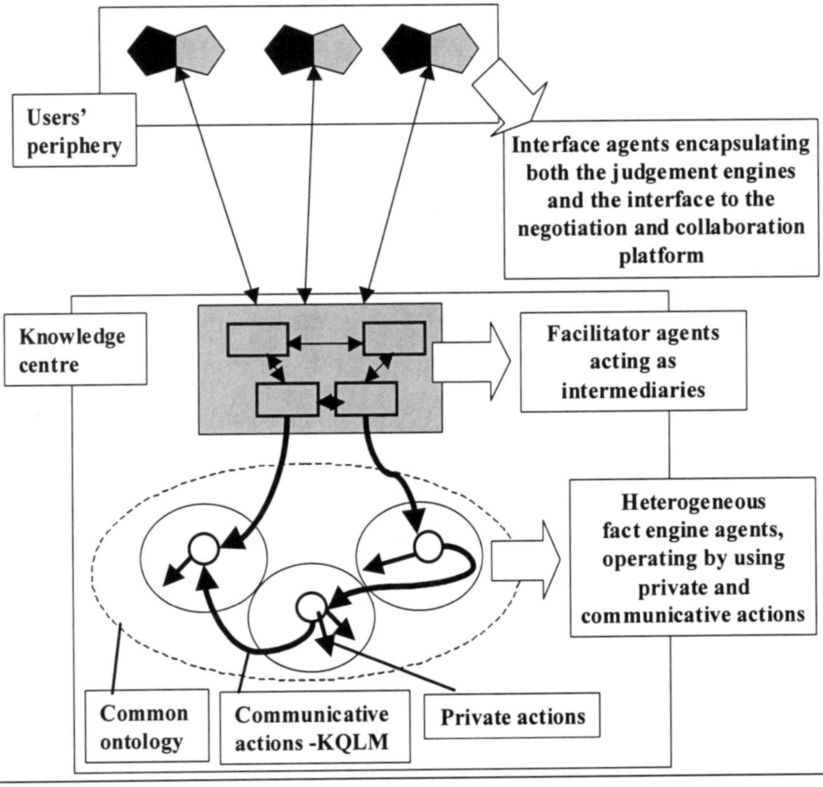

Figure 5.1. Conceptual agent–based architecture for an NDDSS

It needs to be stressed that this is a rather general conceptual architecture which does not touch upon many difficult implementation issues. A concrete NDDSS design and development problem will certainly open up many such implementation issues, and possibly even raise the question of whether the agent oriented approach is suitable at all at the implementation level. Nevertheless, there seem to be a lot of value in adopting this approach at the conceptual level. The essential problem in the design of NDDSSs is the *integration* of the three functional components into one design framework. The tasks that these three components need to perform are very different, and the use of more traditional design methodologies may lead to design specifications that are too complex. This in turn will make the implementation even more difficult. The adoption of an agent-based approach offers the possibility to provide an abstract, high level system specification, which is comparatively less complex, and which emphasises the key integration issues. Once the entire system is decomposed and specified in terms of agents, the implementation issues become relatively easier because each such implementation issue can be approached to a large extent separately for each agent.

In conclusion, it is obvious that fully developed NDDSSs, as they are proposed here, are still some way removed in the future. Their design and development still requires a lot of research, testing and prototyping, most probably with different agent architectures for different decision making problems. It is entirely possible that in some cases the agent-oriented approach may not be required at all. As a *general* approach to NDDSSs design and development, however, the software agent technology clearly has a lot of potential, and its advances elsewhere need to be followed closely and constantly related to the present applications.

Chapter 6

Conclusions

While this study introduces NDDSSs as a viable approach for experiencing and solving the sociotechnical problems of decision making processes in water and environment related problems, it is obvious that their design and development will be by no means simple. From the material presented in all the previous chapters, we may conclude that the task of constructing NDDSSs capable of dealing with, and supporting the multiple aspects of participatory decision making processes is extremely complex. It is in fact clear that it cannot be reduced to one identified problem for which solutions can be sought and proposed. With the primary aim of this study being the basic introduction of this new approach, it can be said that instead of providing a set of such solutions, it actually opens up many more problems. At this early stage, any attempt to provide some definite conclusions in terms of *empirical results* that can be immediately applied in practice is not only difficult, but also quite premature. We can nevertheless try to revise and summarise all the aspects of NDDSSs as they have been presented in this work, while pointing out the most important problems that lie ahead in their future design and development.

A considerable effort in this study went into justifying the introduction of NDDSSs into our societies. The theoretical background presented in Chapter 2, with its broader analysis of our present social condition, was aimed at showing how the very nature of our current problems in general, and of the decision-making problems in particular, requires the adoption of *sociotechnology* as a unifying concept. Many different aspects of our current postmodern condition of society were analysed, and their relations to the immediate tasks of NDDSS design and development were already summarised in the Theses 1-8 of Chapter 2. All these different aspects are at the same time connected to one main idea: that the *social* and the *technical* have nowadays become so intertwined and interdependent, that their separation as two different realms of human thought and action has ceased to be an advantage, but has instead become more of an obstacle. As two separated domains, neither of them is capable of fully accommodating the hybrid nature of our current decision-making problems. The social domain, where indeed the actual decisions appear to be made, is *itself* sustained by technology, and without reliance on technology its processes become paralysed. On the other hand, the disregard for the actual social realities inherent in many, and nowadays prevalent technocratic practices, frequently results in solutions that are scarcely applicable at all in practice, and often detrimental to the actual social environment.

The limitations of such technocratic approaches are most obvious in the fields of water and the environment, where the social context of the decision making problems is characterised by the presence of large numbers of diverse and increasingly complex interests of numerous stakeholders. These interests vary from corporate interests to the interests of individuals, or groups of individuals gathered around common concerns, and all these individuals and groups insist on participating in the decision making processes. The task of satisfying these demands needs to start out from the realisation that the participation of such individuals and groups will become meaningful only if it is a participation based on *knowledge*, and then on knowledge of a considerable variety of kinds. In particular, the separation between narrative knowledge and (modern) scientific knowledge needs to be overcome, and their mutual relation transformed from one of adversity into one of synergy. This realisation is most urgently needed within the context of development in third world societies. At the same time, the ongoing process of knowledge commodification needs to be positively exploited for the satisfaction of the growing knowledge demands, characterised by very diverse knowledge needs. This imperative is nowadays expressed in many areas of business activity, where it is often described as one of 'mass customisation'. Indeed, conferences are now held with just this name. In view of its potentially oxymoronic nature, however, we prefer to avoid this term.

The role of advanced information and communication technologies in the knowledge commodification process is central. From one side it is clear that these technologies can be used to direct the knowledge commodification process in a quite opposite direction than the one proposed here, and thereby increase even further the inequalities in knowledge access. On the other hand, the real potential for growth in the knowledge industry resides in seizing the opportunities of *mass consumption of knowledge* to promote learning processes, both at the individual and at the social level, and indeed at both interdependently. However, the mass consumption of knowledge requires the employment of radically different methods then those employed to promote the mass consumption of other products. The knowledge needs are anything but uniform, and their diversity brings with it the further imperative of employing ICT for providing customised and even personalised knowledge products.

This requirement is closely related to one of the main objectives in the proposed new approach for participatory decision making, which is the direct participation of *individuals* in the decision making processes. If their participation is to become truly active, their role as an *informed general public* is simply not enough. Instead, they need to become *engaged* in the processes as *subjects,* capable of their own judgements and contributing through their own creative thoughts and actions. In this process they need to extend their learning beyond the limits of the problems immediately at hand, with their possible solutions and accompanying consequences. They have to extend their learning to comprehend the *values* of other persons and interest groups, and, also – which is usually the most difficult of all – their own personal values and personal interests. When systems like NDDSSs are described as among the *principal sociotechnical means to promote intersubjectivity*, the emphasis is exactly on their capacity to help the individuals engaged in the process to *change their attitudes* towards their societies and towards nature, and thus to modify their own value systems. Designing systems with such a capacity requires the making of special uses of Internet and ICT technologies, which go much further than the simple provision of personalised content. Instead of 'decision support systems' understood in the traditional sense, these new constructs need to be developed as decision support *environments*, which will enable the participation and active engagement of all interested parties, including concerned individuals.

On the basis of the background analysis presented in Chapter 2, Chapter 3 introduced the concept of NDDSSs, together with their basic design requirements. In order to meet these design requirements, these systems are conceptually separated into two parts, the knowledge centre and the user periphery. The knowledge centre has the role of acquiring and storing the appropriate scientific knowledge and disseminating it in appropriate forms to the user. The user periphery on the other hand has the role of combining the scientific facts with the users' beliefs and attitudes in order to make judgements. Besides this conceptual division, NDDSSs consist of three functional components: the fact engine component, the judgement engine component, and the platform for collaboration and negotiation.

The fact engine component is located in the knowledge centre and is responsible for the production of the (scientific) facts, primarily by providing storage and access to a collection of models and databases related to various aspects of the physical system in question. Because of the envisaged large number of users of a given NDDSS, with potentially very diverse knowledge and data needs, this component will consist of many different kinds of models, including what are commonly called 'data models', such as GIS systems, satellite image processing systems, GPS systems and many others. The primary direction in the design of the fact engine component is towards a more *integrated modelling*. Apart from the various models, different kinds of software components, encapsulating different kinds of fact-providing knowledge (domain expert knowledge, legal or administrative knowledge etc.) may also act as fact engines. The most outstanding problem in the design of the fact engine component, both for the purpose of integrated modelling, and for the efficient integration of various heterogeneous software components, is that of *software interoperability*. The use of already integrated modelling systems as fact engines can greatly reduce this problem, but it will nevertheless remain a central issue in the future design and development of this kind of component. Different decision making problems will involve different tasks and domains, and different technologies for achieving the required interoperability may be employed. One possible *generic* framework for dealing with this problem, based on agent orientation has been proposed in Chapter5.

While the fact engine component provides the *same* facts as *common* to all users/participants, the judgement engine component, which is distributed over the users' periphery, allows the users to formulate *autonomously their own*, *different* judgements about the particular interventions in the system. Although very complex and convoluted, this process can be *described* by adopting a phenomenological viewpoint, which sees all judgements as being mediated by immediate experiences of objects, as *phenomena*. It occurs through the *interaction* of the user's *intentionality network*, consisting of different mental states (beliefs, desires, intentions etc.), with the facts received from the knowledge centre. This descriptive approach can be used to express something of the structure of the judgement forming process, but the actual design of judgement engines needs to rely on general semiotic studies of the actual signification and communication processes through which such interaction results in both immediate and deeper experience. The focus of judgement engine design is therefore in expanding the notion of 'user interface' into a new kind of distributed software component for user interaction, where such immediate experiences can be created through an appropriate use of signs and sign vehicles. These design issues become even more complex when extended with the requirements placed upon the judgement engines to support not only the user's own autonomous judgements, but to assist also the positioning of that user within the social landscape created from the aggregated judgements of all other users/participants. The design of judgement engines is one of the most difficult aspects of NDDSS development. It is conditioned by our improved understanding of judgement formation processes, which needs yet further development of theoretical concepts, as well as the adoption of new kinds of semiotic concepts which were not previously considered as central at all to the development of hydroinformatics systems. This development at the level of theory, proceeding from the most basic level of definitions, is essential within the NDDSSs context, since the distributed judgement engines are in fact critical components through which the users become *engaged* in the whole process. This is currently an ongoing project in hydroinformatics, but it proceeds beyond the more pragmatic interests of the present work.

The platform for negotiation and collaboration can be seen as distributed both over the users' periphery and the knowledge centre. The main role of this component is to *facilitate* the collaboration and negotiation processes. Because of the large number of users, the potentially large number of issues and the great range of possibilities for trade-offs, or compensation agreements, these processes cannot be much automated, and will need to be worked out between the users/participants themselves. The support offered by the platform for collaboration and negotiation starts out by providing a clear presentation of the social landscape to each and every user/participant, and then in such a way that the users can become aware of their own positions relative to other participants, or the whole participating community. At the same time, the platform needs to allow for the *transparent* presentation of the positions of each user/participant. Because the negotiation and collaboration processes are dynamic and non-linear, the platform needs to contain procedures for monitoring and updating the social landscape (carried out predominantly by the components located in the knowledge centre), and for maintaining the required transparency (as achieved mainly through the components distributed at the users' periphery). The main challenge in the design of this component is the extension of the platform's capabilities with components capable of 'intelligent' support to the users in terms of suggesting options with mutual benefits and possibilities for building coalitions or alliances. The details of these aspects were not addressed in this study, but it seems quite obvious that the direction to be followed is towards exploiting the potential of data mining approaches for these purposes.

The *Aquavoice* prototype, presented in Chapter 4, which has been developed within the framework of this study, tries to meet as many of these requirements of an NDDSS as possible. Designed as a multi-player role-play game which can be played over the Internet, it deals with a complex conflict situation involving four institutional (corporate) role players and an unlimited number of role players acting as individually concerned citizens. Although the actual decision problem at hand appears as a relatively simple *choice* problem (among several different scenarios of groundwater abstraction), the conflicting interests of the role players result in a situation which is more typical of an NDDSS approach. The use of a hydrological model constructed with the integrated MIKE SHE modelling system as a fact engine of *Aquavoice* practically eliminates the software interoperability problem in this situation. While only a limited number of MIKE SHE results are actually used in an *Aquavoice* session, the full model has been deployed on the Internet, including the possibility for running remote simulations. This example shows the potential of such applications for fact engine development, and even for more complex decision problems which require the provision of a large number of diverse facts and the generation of alternatives through running additional model simulations.

The judgement engine component of Aquavoice has been developed without resorting to the traditional use of value functions. Instead, the users/participants are provided with an interface through which they can 'see' the facts provided by the MIKE SHE model, and then provide their judgements using linguistic terms. A method for Fuzzy Multiple Alternative Decision Making is then adapted and used to combine the provided judgements and evaluations of the proposed scenarios into particular user's position represented as individual ranking of the scenarios. This methodology has been chosen in order to avoid as much as possible the 'locking' of the users into some predefined system for valuation. This leaves enough freedom to the users/participants to form their autonomous judgements and evaluations. A lot of attention has been put into providing an adequate *transparency* of each user's positions by designing Web-pages which are continuously updated as users interact with the system.

The platform for collaboration and negotiation is the least developed component of *Aquavoice*. A simple aggregation procedure is provided which can serve as basis for initiating the negotiations between the different role players. Each role player can check his or her own position against the positions of the four institutional role players, and against the overall response of the community. As the users carry out their negotiations and collaborations through a chat-like interface, each change in any user's position (in terms of a new judgement or evaluation) is immediately reflected in the overall response of the community, and the new 'state' is 'pushed' to all the users. Similarly to the judgement engine component, this approach is chosen for enhancing the *transparency* of the negotiation and collaboration process.

The integration of all three functional components into one NDDSS framework has been achieved by using various Java technologies. The use of a three-tier concept, which has been followed throughout the *Aquavoice* design, allows initially for an appropriate conceptual and implementation separation, and then for the subsequent integration of all *Aquavoice* components. While the use of Java applets has been very important in constructing the various user interfaces of the system, the most significant advantage of Java came from its powerful networking capabilities for the integration of all system components into one framework.

At the end of Chapter 4, the design of the *Aquavoice* system has been compared to the actual design requirements for an NDDSS presented in Chapter 3. It may be concluded that these requirements were closely followed, and that *Aquavoice* can indeed serve as a proper demonstrator of the basic concepts of an NDDSS. A number of further functional and technical improvements have been proposed, however, which are needed for the fuller development of *Aquavoice* as a multi-player role play game that can be played over Internet. Some of these improvements are in fact indispensable for the proper Internet-based testing of the system by a large number of users.

Although *Aquavoice* may serve as an initial development, which exemplifies the basic concepts of an NDDSS, it is obvious that the elaboration of a fully developed NDDSS still remains a much more complex undertaking. Such a fully developed NDDSS is associated with so many different problems that its design and development becomes extremely difficult without using some more generic conceptualisation framework. Chapter 5 proposes that such a framework could be based on the agent orientation paradigm, with all the systems components, at least, conceptualised as software agents. Within this framework, different kinds of software agents are proposed for the three functional components of an NDDSS. For the fact engine component, the interoperability of the heterogeneous software components in the knowledge centre can be achieved by using software agents operating in a *society* and communicating through a high level Agent Communication Language. The distributed judgement engines, together with the components of the platform for negotiation and collaboration which are distributed at the users' periphery, can be conceptualised as so-called *interface* agents that possess learning capabilities and express adaptive behaviour. These two kinds of agents are interconnected by a number of *facilitator* agents, located at the knowledge centre, which coordinate the communication exchange between the agents and support the negotiation and collaboration process. This approach proposes the conceptualisation of an NDDSS as a *federated* system of agents.

While the origin, the concept and the possible ways forward in developing NDDSSs are rather thoroughly investigated in this thesis, the question of their introduction in practice still remains open. The answer to this question as always, needs to be sought in the interdependence of new technical developments and their corresponding social conditions. As with any other sociotechnical development, in order for NDDSSs to make their way into our societies, three basic conditions have to be satisfied. The first one is the *need* for such a development, in other words, the existence of a certain problematic that necessitates the new development since it cannot be resolved with the existing means. The second condition is related to the existence of enough *resources,* in the broadest sense of this term, in order to achieve that particular development. The third condition may be expressed as the existence of certain *social ethos,* with a sufficient inclination towards the new development and a readiness to take up its support. The *need* for systems like NDDSSs has hopefully become quite apparent from this thesis. Although there is still a need for a further development of theoretical concepts related to NDDSSs, the existing human and technological resources are certainly enough for beginning the development of NDDSS-like systems. In particular, the advanced enabling technologies in terms of ICT are already present, as exemplified by the development of the *Aquavoice* prototype.

The question of introduction of NDDSS-like systems into our societies, even if they are further developed with a high level of sophistication, may depend critically on the third condition. In concluding this chapter we will briefly look at the possible ways in which a social ethos that is supportive of this new approach may emerge. Because this approach involves some radical changes in the established order of decision making, which means that they are aimed at transforming the existing knowledge/power relations in our societies, the ways in which they can find their social support does not seem to be straightforward. We will propose here three potential approaches through which this can be achieved.

1) The first approach is the continuation of development of Internet distributed multi-player games, of the *Aquavoice* type, which will in future include more NDDSSs features. This strategy makes use of the central role of play existing in virtually all human societies, and indeed in most societies of mammalians. Although it is quite difficult to explain the relation between play and socialisation in terms of some causal relation, playing games has been an indispensable activity of people throughout history, and one in which each culture finds its manifestation. From one side, playing games may appear as a 'purposeless' activity, which lacks seriousness, and needs to be separated and even prohibited from some environments that are supposed to be dedicated to activities considered as 'real work'. With the background presented in Chapter 2, however, it is relatively easy to locate the origins of this particular dichotomy within the modern constitution, and even to relate it to the separation between scientific and narrative knowledge. The opposite view, which can be associated with our present postmodern condition of society, recognises the role of play in human culture and objects to its devaluation. However, the relation between games and society remains quite highly dualistic in another sense, which can be associated with the duality of our postmodern / hypermodern condition. The postmodern stance sees games as epitomising those attitudes which are so much needed in our societies, such as *free* engagement, daring, cooperation, or healthy competition, and seeks various modes through which game playing can be used for transferring these attitudes into other, more 'serious', domains of human thought and activity. On the other hand, the hypermodern society *uses* games as one of the primary media through which it reasserts and sustains its own cultural and ethical values. As we have seen in Chapter 2 (see Thesis 5), this approach may take sophisticated paralogical forms whereby some computer and video games become crucial components for the creation of various simulacra, in which alone many persons can find an illusion of satisfaction and safety.

This awareness of this situation will help in determining the strategies that can be employed for further development of NDDSS-like games. Because the main idea behind such games is that they maintain a strong connection with actual 'real-world' water and environmental problems, it seems that such developments must be initially promoted by organisations traditionally involved in the construction of real-world hydroinformatics systems, and then in particular from their modelling departments. Specialists working in these organisations have contacts with many existing problems which require the NDDSS-approach, and are familiar both with the appropriate modelling techniques and with the enabling ICT technologies. On the other hand, because of their lack of expertise in providing appropriate entertainment features, they will probably have to make strategic alliances with organisations belonging to the much larger computer-gaming industry. Such organisations are continuously looking for new ideas for game development. Some of them may also find in such an alliance an appropriate response to the increasing demands directed towards this industry to show more social responsibility in the design and developments of their new products. The most important issue in such an alliance, and probably the one where most tensions might arise, will be the right balance between 'real-world content' of the game and its entertaining features. At the moment of this writing, this seems like a rather unlikely scenario, even if the potential for such developments already exist. With the right initiatives, however, which should again come from the designers of real-world hydroinformatics systems, such developments could be materialised.

The basic assumption of this approach is that through the penetration of these NDDSS-like games into the culture of our societies, the awareness of the ordinary people about the possibilities of this new approach will be raised. Once this awareness is raised to a sufficient level, the pressure on the existing government and market institutions for extending this approach to actual real-world problems will naturally increase.

2) The second class of environments in which the social ethos in support of the NDDSS systems could arise is that of various governmental and non-governmental organisations (NGOs). In most countries, water- and environment-related problems fall mainly under governmental jurisdiction, even if they involve many other, and different, interests. This makes the governmental organisations an obvious place from where NDDSSs can be disseminated. This opportunity needs to be seen within the context of the changing status of governments, both at the national and at the local level. The tendency of transferring governmental control in various domains to the private sector, and to 'self regulating' market mechanisms, continuously reduces the power of governments. At the same time, in many instances, the complacency of governments participating in this process is often perceived by their electoral constituency as some sort of 'betrayal'. Water is still very much associated with the notion of 'public good', and governments have a primary responsibility for its safeguarding. The situation with the environment is similar. The various forms of privatisation, which enter in these areas with governmental consent, are far from being always welcomed. As a consequence, state and local governments are increasingly losing not only their power, but also their *legitimacy*.

This situation contributes strongly to the general movement away from representative democracy and towards participatory democracy. The era of mass membership into political parties which provide links between elected government officials and their electorate, is declining. In order for governments to continue playing their important role they need to take actions for embracing the new reality, and support the emerging tendencies towards more direct and participatory decision making. These tendencies are essentially sociotechnical in their nature, being strongly catalysed by the Internet and the rapidly advancing developments in mobile telephony. Making the introduction of NDDSSs into society their own 'business', is one action that governments may take that would have the potential to restore their otherwise eroding legitimacy.

As a parallel response to this situation, we witness the extraordinary ascent of various NGOs. These organisations have managed to fill the empty spaces for social action abandoned by the governments, and they now attract a broad popular appeal. The growth of these organisations has been such that their conglomeration is nowadays often associated with the term 'civil society'. Such organisations seem like rather natural supporters of the NDDSS approach for water and environmental problems. The very fact, however, that they are not democratically elected organisations, constantly brings their legitimacy into question, and because of this their power is in most cases rather restricted, even as their freedom of action is so much less restricted.

In a situation where governments and NGOs each have what the other one lacks, many voices nowadays argue for their mutual partnership. For such partnerships to succeed, however, both have to demonstrate their legitimacy. Promoting systems like NDDSSs into society is the kind of common project on which they can join their forces to achieve that purpose.

3) The third approach that can be sought for the introduction of NDDSSs into our societies proceeds directly through the market. It is an approach that looks directly at the opportunities offered by the processes of knowledge commodification. It is very often stressed how these processes are driven by technology, such as continuously improving Internet solutions, or the introduction of third-generation (3G) telephone networks. At the same time, this process of constant introduction of new, more efficient, technologies is paralleled by the lack of appropriate and useful 'content' that is to be delivered through them. While the potential for delivering customised and personalised content using these technologies is repeatedly used in the advertising campaigns for these technologies, their market success depend critically on the question of 'content'. Water and environment related problems are certainly one area where fast and always-available personalised content delivery is needed. There is a lot of potential in NDDSSs-like systems for development of new business opportunities. The initiatives for such businesses need to come again primarily from organisations involved in the development of hydroinformatics systems, but now the strategic alliances need to be sought with market-oriented companies that are promoting new technologies and are searching for appropriate and useful content.

Although this approach of 'making business out of individual stakeholder participation' seems to have great potential, its realisation is clearly not without obstacles. The market sector of the new technologies does not yet perceive the area of water as sufficiently profitable. During the '.com' boom on the financial markets there were simply no real incentives for this sector to look for serious applications in other areas. With the end of the so-called 'rush to the market', this situation will hopefully change. Another obstacle to this approach is precisely the fact the most water and environment-related problems still fall under government responsibility, and developing business opportunities in this sector without the involvement of governmental institutions seems to be very difficult. This situation again shows that the introduction of NDDSS-like systems into our societies requires new kinds of strategic partnerships.

The three proposed approaches only point towards the potential areas from where the introduction of NDDSSs into our societies may be taken up. It is not at all clear which of these approaches, or their combinations may succeed. Moreover, it is most likely that different decision making contexts, and especially different societies and cultures, will require different approaches. This discussion can nevertheless serve as an initial guideline for directing future developments of systems like NDDSSs.

At the end of this discussion it should become clear that the field of hydroinformatics, and its practitioners, should take the primary responsibility for the introduction of NDDSSs into our societies. The responsibility of hydroinformatics cannot be limited to the analysis, design and development of these systems, and needs to be extended with these crucial aspects of their promotion into social practice. This is what makes the whole study of NDDSSs a truly hydroinformatics study. With this realisation, we now need to revisit the discussion presented in Chapter 2 (section 2.8) about the *middle ground* that hydroinformatics aims to occupy. We have presented this middle ground of hydroinformatics in Figure 1.1 as the centre of gravity of the triangle formed by three poles: water-related physical sciences, information and communication technology, and the awareness of the social context where hydroinformatics systems are being introduced. The study of NDDSSs, as has been presented in this thesis, leads us to extend this representation into a similar, but more expressive form, given in Figure 6.1.

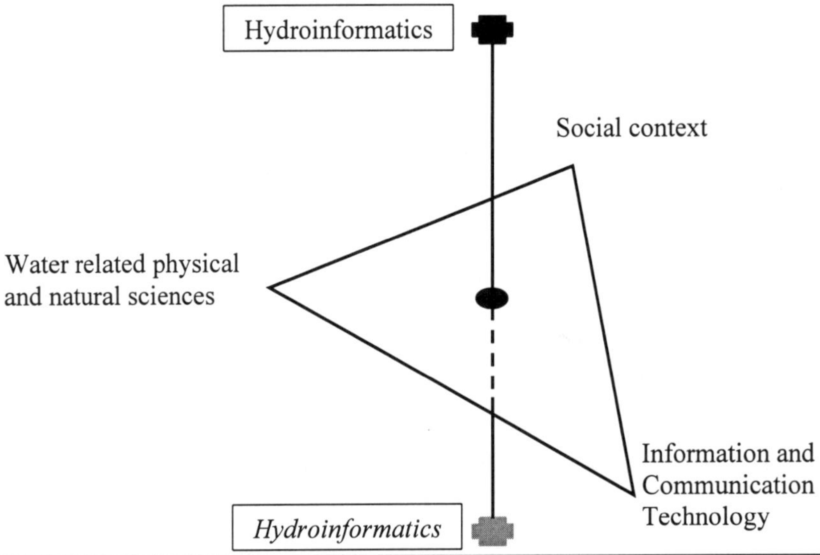

Figure 6.1. The 'fluctuating' middle ground of hydroinformatics

While the three poles remain in principle unchanged, we can now observe the necessarily 'fluctuating' nature of hydroinformatics, above and below the triangle. Most of the time, hydroinformaticians see the 'self-aware' position of hydroinformatics, which is above the triangle, from where it 'oversees' the developments in each of the three areas, and draws upon all of them for its own endeavours. When we define *hydroinformatics as sociotechnology* we must realise that it is *itself* shaped, directed and transformed by the various developments in the three respective areas. At times, we need to step aside and look at its position below the triangle, even if only in order to be able to restore its position above it. For only from this position of 'self-awareness' will it be capable of achieving its own mission of serving humanity and our planet. What I sincerely hope will be retained by the readers of this work is that the achievement of this mission remains, as always, our own, human, responsibility.

References

Abbott, M.B., Bathurst J.C., Cunge J.A., O'Connell P.E. and Rasmussen J., 1986a, An Introduction to the European Hydrological System – Système Hydrologique Européen 'SHE', 1: History and philosophy of a physically based distributed modelling system. *Journal of Hydrology*, 87, 45-59.

Abbott, M.B., Bathurst J.C., Cunge J.A., O'Connell P.E. and Rasmussen J., 1986b, An Introduction to the European Hydrological System – Système Hydrologique Européen 'SHE', 2: Structure of a physically based distributed modelling system. *Journal of Hydrology*, 87, 61-77.

Abbott, M.B., 1991, Hydroinformatics: Information Technology and the Aquatic Environment, Aldershot, UK/Brookfield, USA: Ashgate.

Abbott, M.B., 1992, The theory of the hydrological model, or: The struggle for the soul of hydrology, in *Advances in Theoretical Hydrology: a tribute to Jim Dooge*, ed. O'Kane P., pp.237-254, Elsevier, Amsterdam.

Abbott, M.B., 1993, The electronic encapsulation of knowledge in hydraulics, hydrology and water resources, *Adv. Wat. Resources*, 16, pp 22-39.

Abbott, M.B., 1994, Hydroinformatics, a Copernican revolution in hydraulics, *Journal of Hydraulic Research*, 32, Extra Issue, pp. 3-14.

Abbott, M.B., 1996. The sociotechnical dimension of hydroinformatics, *Hydroinformatics '96, Müller(ed.)*, Balkema, Rotterdam,

Abbott, M.B., 1998. Future business opportunities: Future human responsibilities, *Hydroinformatics '98, Babovic & Larsen (eds)*, Balkema, Rotterdam,

Abbott, M.B. and Jonoski, A., 1998. Promoting collaborative decision-making through electronic networking, *Hydroinformatics '98, Babovic & Larsen (eds)*, Balkema, Rotterdam,

Abbott, M.B., 1999a, Introducing hydroinformatics, *Journal of hydroinformatics*, 01.1, pp. 3-19,.

Abbott, M.B., 1999b, Forchheimer and Schoklitsch: a postmodern retrospection, in *Proceedings of XXVIII IAHR Congress*, pp. 135-154, Graz, Austria (ed. Graf, W.H.)

Abbott, M.B., 2000a,The gender issue in hydroinformatics, or Orpheus in the Underworld, *Journal of hydroinformatics*, 02.2, pp.88-104.

Abbott, M.B., 2000b, Applications of Meinongian object / value theory in hydroinformatics, In 4th *International Conference Hydroinformatics 2000,* Cedar Rapids, IA, USA.

Abbott, M.B., 2001, The democratisation of decision-making processes in the water sector I, *Journal of Hydroinformatics* 03. 1, pp.23-34.

Abbott, M.B. and Jonoski A., 2001, The democratisation of decision-making processes in the water sector II, *Journal of Hydroinformatics* 03. 1, pp.35-48.

Abbott, M.B., Babovic, V. and Cunge, J., 2001, Towards the Hydraulics of the Hydroinformatics Era, *Journal of Hydraulic Research*, 38, 4, pp. 339-349.

Adams, P., 1994, Property Rights and Bioregionalism, CATO Policy Report, Nov/Dec., CATO Institute, Washington D.C.

Appignanesi R. & Garratt C., 1995, *Postmodernism for beginners.* Icon books Ltd. Cambridge.

Armstrong, S.J. & Botzler, R.G., 1993, *Environmental Ethics: Divergence and Convergence*, McGraw-Hill, Inc. New York.

Axelrod, R., 1984, *The evolution of cooperation*, Basic Books, New York.

Babovic, V. and Keijzer, M., 2000., Genetic programming as a model induction engine, *Journal of Hydroinformatics*, 02, 1, pp. 35-60.

Baudrillard, J.,1985. The Ecstasy of Communication, in *Postmodern Culture, ed. Foster. H.*, Pluto Press, London.

Bauman, Z., 1996, *Postmodern ethics,* Blackwell Publishers Inc., Oxford.

Botella, L. (1995). *Personal construct psychology, constructivism, and postmodern thought.* In R.A. Neimeyer & G.J. Neimeyer (Eds.), Advances in Personal Construct Psychology (Vol. 3) (pp. 3-36). Greenwich, CN: JAI Press

Bregman, J.I. and Mackenthun K.M., *Environmental Impact Statements*, Lewis Publishers, London.

Brooks R.A., Breazeal C., Marjanovic M., Scassellati B. and Williamson MM., 1999, The Cog Project: Building a Humanoid Robot, in *Computation for Metaphors, Analogy, and Agents. Lecture notes in Artificial Intelligence,* a subseries of *Lecture Notes in Computer Science*, Springer-Verlag.

Bruin J.A. and Heuvelhof E.F., 1999, Scientific expertise in complex decision making processes. *Science and Policy, special issue of the Journal of the International Science Policy Foundation*, Vol. 26, No. 3, pp.179-192, Beech Tree Publishing.

Cairncross, F., 1997, *Death of Distance*, Harvard Business School Press, Boston Massachusetts.

Callon, M. 1986. Some elements of a sociology of translation; domestication of the scallops and the fishermen of St. Brieuc Bay. In *Power, Action and Beliefs: a New Sociology of Knowledge,* J. Law (ed.), London: Routledge.

Castells, M., 1996, *The Rise of the Network Society*, Blackwell Publishers Inc., Oxford.

Castells M., 1997, *The Power of Identity*, Blackwell Publishers Inc., Oxford.

Castells M., 1998, *End of Millennium*, Blackwell Publishers Inc., Oxford.

Chen, S. and Hwang, 1992, C, Fuzzy Multiple Attribute Decision Making, *Lecture Notes in Economics and Mathematical Systems*, Springer-Verlag,.

Daly, E. H., 1997, Science, Religion and Sustainable Development, in *Revelation and the Environment*, pp. 62-66, (editors, Hobson S, and Lubchenco J), World Scientific Publishing, London.

Davignon, R, 1985, *Le mal chez Gabriel Marcel*, Recherches, nouvelle série – 4, Éditions Bellarmin, Montreal.

Devlin, K., 1995, *Logic and Information*, Cambridge University Press.

DHI, 1993, *MIKE SHE WM, Release 5.1, Water Movement Module. A Short Description*, Hørsholm, Denmark.

DHI, 1997, *Mike Impact User Guide*, Hørsholm, Denmark.

DHI, 1998, *MIKE SHE, Water Movement Module (v. 5.3), User Guide and Technical Reference Manual*, Hørsholm, Denmark.

Dibike Y.B., Solomatine, D., and Abbott M.B., 1999, On the encapsulation of numerical-hydraulic models in artificial neural network, *Journal of Hydraulic Research*, 37, 2, pp.147-161.

Dignum F. and Greaves M., 2000, *Issues in Agent Communication*, (eds.) *Lecture notes in Artificial Intelligence,* a subseries of *Lecture Notes in Computer Science*, Springer-Verlag.

Drucker, P.F., 1994, The Age of Social Transformation, *The Atlantic Monthly*, Volume 274, No. 5; pp. 53-80.

Duhem, P. 1913-1959, *Le système du monde. Histoire des doctrines cosmologiques de Platon a Copernic*, (in 10 volumes), Herman Paris.

Duhem, P. 1985. Medieval cosmology; theories of infinity, place void and the plurality of worlds, ed. And trans. Ariew, R., University of Chicago, Chicago, Ill.

Dunn J., 2000, *The Cunning of Unreason*, HarperCollins*Publishers*, London.

Durning D., 1999, The Transition from Traditional to Postpositivist Policy Analysis: A role for Q-Methodology. *Journal of Policy Analysis and Management*, Vol. 18, No.3, pp. 389-410, John Wiley & Sons Inc.

Dyson, E, Gilder, G, Keyworth, G.A., and Alvin Toffler, 1994, Cyberspace and the American Dream: A Magna Carta for the Knowledge Age, *Future Insight* 1.2.

Easterly, W., 2001, *The Elusive Quest for Growth*, MIT Press.

Eco, U., 1976, *A theory of semiotics*, Indiana Univ., Bloomington, USA.

Edmonds, B. and Dautenhahn, K. (eds.), 2001, Starting from Society: the Application of Social Analogies to Computational Systems, *Special Issue of Journal of Artificial Societies and Social Simulation* Vol. 4, no. 1.

Ellul J., 1964, *The Technological Society*, Vintage Books, New York.

European Environmental Bureau, 1998, *Doors to Democracy: Current Trends and Practices in Public participation in Environmental Decision Making in Western Europe*, European Environmental Bureau.

Findley J.N., 1995, *Meinong's Theory of Objects and Values*, second edition, Gregg Revivals.

Finin, T., Labrou, Y., and Mayfield, J., 1997, KQML as an Agent Communication Language, in *Software Agents*, J.M. Bradshaw (ed.), MIT Press.

Fisher, R. and Ury W., 1987, *Getting to Yes – Negotiating Agreement Without Giving In*, Arrow Books, London.

Franklin, S. & Graesser, A., 1996, Is it an Agent, or just a Program?: A taxonomy for Autonomous Agents, in *Proceedings of the Third International Workshop on Agent Theories, Architectures and Languages*, Springer-Verlag.

Fukuyama, F, 1989, The End of History, *The National Interest,* 16, pp.3-18.

Fukuyama, F, 1992, *The End of History and the Last Man*, Penguin Books, UK.

Gallagher, J., 1998, Dostoevsky as Philosopher, Guest Lecture at the Philosophy department of UC Davis; *(http://hume.ucdavis.edu//phi151/NOV28LEC.HTM)*

Genesereth, M.R., 1997, An Agent Based Framework for Interoperability, in *Software Agents*, J.M. Bradshaw (ed.), MIT Press.

Gergen, K.J. and Thatchenkery T, 1997 Organizational science in a postmodern context. *Journal of Applied Behavioral Science*. 32, 356-377. .

Gilbert, N., 1996, *Simulation: a emergent perspective*, conference on New Techologies in Social Sciences, Bournemouth, UK, and LAFORIA, Paris, France *(http://www.soc.surrey.ac.uk/research/simsoc/tutorial.html)*

Gruber, T.,1993, A Translation Approach to Portable Ontology Specifications, *Knowledge Acquisition*, 5(2):199-220.

Habermas. J., 1985, Modernity – An Incomplete Project, in *Postmodern Culture, ed. Foster. H.*, Pluto Press, London.

Haywood, T., 1995. *Info-Rich – Info-poor; Access and Exchange in the Global Information Society*, Bowker-Saur, London

Heidegger., M., 1927 *Sein und Zeit*. Neimeyer, Tuebingen//1962 *Being and Time* (trans. Macquattie, J. & Robinson, E.). Blackwell, Oxford.

Heidegger., M., 1963, *Die Technik, und Die Kehre*. Neske, Pfullingen //1977 *The Question Concerning Technology and Other Essays* (trans. Lovett, W.). Harper, New York.

Hobbes, T., 1651 // 1968, *Leviathan*, Penguin Books Ltd., Harmondsworth, England.

Hobson, S. and Lubchenco, J., (eds.), 1997, *Revelation and the Environment AD 95-1995* World Scientific Publishing, London.

Hoppe R., 1999., Policy analysis, science and politics: from 'speaking truth to power' to 'making sense together'. *Science and Policy, special issue of the Journal of the*

International Science Policy Foundation, Vol. 26, No. 3, pp.201-210, Beech Tree Publishing.

Horkheimer, M. and Adorno, W., 1944, *Dialektik der Aufklarung*, Social Studies Associations New York, // 1969, S. Fischer Verlag GmbH, Frakfurt am Main // 1973, *Dialectic of the Enlightenment*, trans. John Cumming, Lowe & Brydone Ltd., Thetford, Norfolk

Husserl, E., 1948, *Erfahrung und Urteil: Untersuchung zur Genealogie der Logik*, Claassen and Goverts, Hamburg //1973, *Experience and Judgement, Investigations in a Genealogy of Logic,* trans. Churchill, J.S. & Ameriks, K., Routledge, London.

Hwang, C. and Masud, A.S.M., 1979, *Multiple Objective Decision Making – Methods and Applications: A State-of-the-Art Survey*, Springer-Verlag, Berlin/Heidelberg/New York.

Hwang, C., and Yoon, K., 1981, Multiple Attribute Decision Making, methods and applications, *Lecture Notes in Economics and Mathematical Systems*, Springer-Verlag,.

Hwang, C., and Lin, M., 1987, Group Decision Making under Multiple Criteria, *Lecture Notes in Economics and Mathematical Systems*, Springer-Verlag,.

Jameson, F, 1985, Postmodernism and Consumer Society, in *Postmodern Culture, ed. Foster. H.*, Pluto Press, London.

Jameson, F, 1991, *Postmodernism, or, The Cultural Logic of Late Capitalism*, Verso, New York.

Jankowski P., 2000, Collaborative spatial decision making in environmental restoration management: an experimental approach, *Journal of Hydroinformatics*, 02, 3, pp. 197-206.

Jennings N.R. and Wooldridge M. J., 1998, *Applications if Intelligent Agents, in Agent Technology Foundations, Applications and Markets, Jennings N.R. and Wooldridge M. J. (eds),* Springer-Verlag.

Jonoski, A. and Abbott, M.B., 1998, Network distributed decision support systems as multi-agent constructs, *Hydroinformatics '98, Babovic & Larsen (eds)*, Balkema, Rotterdam,.

Jonoski, A., 1999, Towards the development of distributed decision support tools for use over electronic networks, *Proceedings of XXVIII IAHR Congress*, Graz, Austria.

Jonoski, A., 2000, AquaVoice: A prototype for internet distributed decision support system, In *4th International Conference Hydroinformatics 2000,* Cedar Rapids, IA, USA.

Karliner, J., 1997, *The Corporate Planet; Ecology and Politics in the age of Globalisation*, Sierra Club Books, San Francisco.

Kaiser-Hill Company, 2001, Model Code and Scenario Selection Report Site-Wide Water Balance Rocky Flats Environmental Technology Site, Report No. 01-RF-0037. (*http://www.dhisoftware.com/mikeshe/Download/RFETS_2-20-01.PDF*)

Kelly J.S., 1978, *Arrow Impossibility Theorems*, Academic Press, New York.

Kimball, R., 1992, Francis Fukuyama & the End of History, *The New Criterion*, Vol.10, No.6.

King, K. and McGrath, S. 2000, Becoming a Knowledge Bank? The World Bank's Emerging Approach to Knowledge, Partnership and Development in the Time of Globalisation, *Learning to make policy - project paper*, University of Edinburgh, (*http://www.ed.ac.uk/~centas/fgpapers.html*).

Klinkenberg, J-M, 1996., *Précis de sémiotique générale*, De Boeck & Larcier S.A.

Kristensen, K.J. and Jensen S. E., 1975, A model for estimating actual evapotranspiration from potential evapotranspiration, *Nordic Hydrology*, 6, 70 – 88.

Lahar, S., 1996, Ecofeminist Theory and Grassroots Politics, in *Environmental Ethics: Divergence and Convergence, Armstrong, S.J. & Botzler, R.G. eds.*,pp. 444-453, McGraw-Hill, Inc. New York.

Latour, B., 1987. *Science in Action: How to Follow Scientists and Engineers through Society*, Cambridge, Mass: Harvard University.

Latour, B., 1993. *We Have Never Been Modern*, London: Harvester.

Latour, B., 1991. Technology is society made durable. In *A Sociology of Monsters: Essays on Power, Technology and Domination.* J. Law (ed.), London: Routledge.

Law, J., 1991. Introduction: monsters, machines and sociotechnical relations. In *A Sociology of Monsters: Essays on Power, Technology and Domination.* J. Law (ed.), London: Routledge.

Lyotard, J-F.,1979, *La condition postmoderne: raport sur le savoir*, Minuit, Paris // 1984 *The postmodern condition: a report on knowledge*, Manchester University Press.

Maes, P., 1994, Modeling Adaptive Autonomous Agents *Artificial Life Journal*, C. Langton (ed.), Vol. 1, No. 1 & 2, pp. 135-162, MIT Press.

Majone G., and Quade E.S., 1980, *Pitfalls of Analysis*, International Series on Applied System Analysis, John Wiley & Sons Inc.

Marcel, G.,1956//1984, *The Philosophy of Existentialism,* Philosophical Library //1995, Carol Publishing Group Edition, New York.

Margolin, V.,1995, The Politics of the Artificial, *Leonardo*, Vol 28, No.5, MIT Press.

Martin J., and O'Kane, J. P, 2000, Development of a high resolution hydrodynamic flood mapping model using one-dimensional hydraulic models integrated with high resolution digital elevation models, In *4th International Conference Hydroinformatics 2000,* Cedar Rapids, IA, USA.

McCarthy, J., 1979, Ascribing Mental Qualities to Machines// 1990 reprinted in *Formalization of common sense,* papers by John McCarthy, V. Lifschitz (ed.), Ablex.

McElroy M., 2000., The New Knowledge Management, *Knowledge and Innovation*, 1, 1, pp.43-67

Miller, H., 1995, The Social Psychology of Objects, conference: *Understanding the Social World*, Huddersfield Univ. *(http://www.ntu.ac.uk/soc/psych/miller/objects.htm)*

Minns A.W., 2000, Subsymbolic methods for data mining in hydraulic engineering, *Journal of Hydroinformatics* 02, 1, pp. 3-13.

Mitcham, C, 1994, *Thinking through Technology: The Path Between Engineering and Philosophy*, University of Chicago Press.

Nasr, S.H., 1999, *The Spiritual and Religious Dimensions of the Environmental Crisis*, Temenos Academy, London.

Neal, A., 1995, The Promise and Practice of Deconstruction, *Canadian Journal of History*, 30, pp. 49-76.

Nehaniv, C.L., 1999, The Second Person – Meaning and Metaphors, in *Computation for Metaphors, Analogy, and Agents. Lecture notes in Artificial Intelligence,* a subseries of *Lecture Notes in Computer Science*, Springer-Verlag.

Nelson R.H., 1997, Does 'Existence Value' Exist?, *The Independent Review*; Spring97.

Norris, C., 1993. *The Truth About Postmodernism*, Blackwell, Oxford.

Norris, C. and Benjamin A., 1988, *What is Deconstruction?*, Academy Editions, London.

Nwana, H.S, 1996, Software Agents: An Overview, in *Knowledge Engineering Review*, Vol.11, No.3, pp.205-244.

Nwana, H.S. and Ndumu D.T., 1999, A Perspective on Software Agents Research, *Knowledge Engineering Review* 14 (2), pp. 1-18.

Owens, C., 1985, The Discourse of Others: Feminists and Postmodernism, in *Postmodern Culture, ed. Foster. H.*, Pluto Press, London.

Palerm J.R., 1999, Public Participation in Environmental Decision making: Examining the Aarhus Convention, *Journal of Environmental Assessment Policy and Management*, Vol. 1, No.2, pp. 229-244, Imperial College Press.

Pawelas A. and Mostert E., Deciphering EIA: An Introduction to Environmental Impact Assessment, *RBA Series on River basin Administration, Working Paper No. 23*, Delft University of Technology, Delft, The Netherlands.

Powel, J., 1997, *Derrida for Beginners*, Writers And Readers Publishing, New York.

Raiffa H., 1982, *The Art and Science of Negotiation*, Belknap, Harward University Press.

Refsgaard J.C and Storm B., 1995, MIKE SHE. in *Computer models of Watershed Hydrology*, V.P. Singh (ed.), pp. 809-847, Water Resources Publications.

Refsgaards J.C., 1996, Terminology, modelling protocol and classification of hydrological model codes. in *Distributed Hydrological Modelling*, M.B. Abbott and J.C. Refsgaard (eds.), pp. 17-39, Kluwer Academic Publishers, The Netherlands.

Roy, A., 1999 *The Cost of Living.* Flamingo, HarperCollins*Publishers*, London.

Russell, S.J. & Norvig, P.,1995, *Artificial Intelligence: A Modern Approach*, Prentice Hall, New Jersey.

Sadler, B., 1996, International Study of the Effectiveness of Environmental Assessment: Final Report - Environmental Assessment in a Changing World: Evaluating Practice to Improve Performance (*http://www.ea.gov.au/assessments/eianet/eastudy/final/index.html*).

Said, E.W., 1985, Opponents, Audiences, Constituencies and Community, in *Postmodern Culture, ed. Foster. H.*, Pluto Press, London.

Searle J.R., 1983, *Intentionality*, Cambridge University Press.

Shiva, V., 1996, Women in Nature, in *Environmental Ethics: Divergence and Convergence, Armstrong, S.J. & Botzler, R.G. eds.*, pp. 458-466, McGraw-Hill, Inc. New York.

Shoham, Y., 1994, Logics of Mental Attitudes in AI: a very preliminary survey, in *Foundations of Knowledge Representation and Reasoning*, G. Lakemeyer and B. Nebel (eds.), Springer-Verlag.

Shoham, Y., 1997, An Overview of Agent-Oriented Programming, in *Software Agents*, J.M. Bradshaw (ed.), MIT Press.

Singh, M. P., 1994, Multiagent Systems: A Theoretical Framework for Intentions, Know-How, and Communications, *Lecture notes in Artificial Intelligence,* a subseries of *Lecture Notes in Computer Science*, Springer-Verlag.

Spinoza, B. de, 1677, *Ethica, //*1983, *Etika*, (in serbo-croat,), transl. Atanasijevic K., BIGZ Beograd.

Spretnak, C., 1996, States of Grace, in *Environmental Ethics: Divergence and Convergence, Armstrong, S.J. & Botzler, R.G. eds.*, pp. 466-473, McGraw-Hill, Inc. New York.

Stauth, R., Sowmann, M., and Grindley, S., 1993. The panel evaluation method: an approach to evaluating controversial resource allocation proposals. *Environmental Impact Assessment Review*. 13 (10):13-35.

Storm, B., and Refsgaard, A., 1996, Distributed physically-based modelling of the entire land phase of the hydrological cycle, in *Distributed Hydrological Modelling*, M.B. Abbott and J.C. Refsgaard (eds.), pp. 55-69, Kluwer Academic Publishers, The Netherlands.

The Regional Environmental Center for Central and Eastern Europe, 1998, *Doors to Democracy: A Pan-European Assessment of Current Trends and Practices in Public participation in Environmental Matters*, The Regional Environmental Center for Central and Eastern Europe.

Thissen W.A.H. and Twaalfhoven P.G.J., 2001, Towards a conceptual structure for evaluating policy analytic activities, *European Journal of Operational Research*, 129, pp. 627-649, Elsevier.

Tuthill R.D. Jr, Petrich C.R., Ciscell M.H., Kissinger B.A. and Oakleaf J.R., 2000, Migration from tabular to spatial data analysis techniques for water management in Idaho, *Journal of Hydroinformatics*, 02, 3, pp. 183-195.

Velickov, S., Price, R. K., and Solomatine, D.P., 1998, Using Internet for Hydroinformatics – Practical examples of client-server modelling, *Hydroinformatics '98, Babovic & Larsen (eds)*, Balkema, Rotterdam,.

Velickov, S., and Solomatine., D.P., 2000, Predictive Data Mining: Practical Examples, in *Artificial Intelligence Methods in Civil Engineering Applications*, *Schleider O.H. and Zijderveld A. (eds)*, Institut fuer Buinformatik, Brandenburg, Gernmany.

Warren, K.J., 1996, Ecological Feminist Philosophies: An Overview of the Issues, in *Ecological Feminist Philosophies, Warren K. ed.*,pp. i-xxvi, Indiana University Press.

Wathern, P., 1992, *Environmental Impact Assessment*, Routledge, London.

Webler, I., Kastenholz, H., and Renn, O., 1995. Public participation in impact assessment: a social learning perspective, *Environmental Impact Assessment Review*. 15 (5): 443-464.

Weingart P., 1999, Scientific expertise and political accountability: paradoxes of science and politics. *Science and Policy, special issue of the Journal of the International Science Policy Foundation*, Vol. 26, No. 3, pp.151-161, Beech Tree Publishing.

Weiss G., 1999, *Multiagent Systems. A Modern Approach to Distributed Artificial Intelligence*, MIT Press.

White, J., 1997, Mobile Agents, in *Software Agents*, J.M. Bradshaw (ed.), MIT Press.

White L.G., 1994, Policy Analysis as Discourse, *Journal of Policy Analysis and Management*, Vol. 13, No.3, pp. 506-525, John Wiley & Sons Inc.

Wildavsky, A., 1993, *Speaking Truth to Power, The Art and Craft of Policy Analysis*, Transaction Publishers, New Brunswick, New Lersey.

Wittgenstein., L., (1953//1992) *Philosophical Investigations*, 3rd ed., translation from German by Anscombe, G.E.M., Blackwell, Oxford.

Woods, B., 1993, *Communication, Technology, and the Development of People*, London: Routledge.

Wolff, M., 1998, *Burn Rate, or How I Survived the Formative Years of the Internet. Weidenfeld*, London.

Wooldridge, M. & Jennings, N.R., 1995, Intelligent Agents: Theory and Practice, *Knowledge Engineering Review,* Vol. 10, No.2.

World Bank, 1993, *Water Resources Management*, A World Bank Policy Paper, World Bank, Washington.

World Bank, 1999, *World Development Report 1998/1999: Knowledge for Development*, for the World Bank published by Oxford University Press.

World Bank, 2001, *World Development Report 2000/2001: Attacking Poverty*, for the World Bank published by Oxford University Press.

World Commission on Dams., 2000, *Dams and Development: A New Framework for Decision Making*, for the World Commission on Dams, published by Earthscan, London.

Yan, H., Solomatine D.P., Velickov, S. and Abbott, M.B., 1999, Distributed environmental impact assessment using internet, *Journal of hydroinformatics*, 01.1, pp.59-70,.

Yunus, M., 1993//1997, *Raising the Productivity of the Poor: Grameen Bank Experience in managing New Technologies.* Reprinted by Grameen Bank, Dhaka.

Yunus, M., 1997, *Towards Creating a Poverty-Free World*, reprinted by Grameen Bank, Dhaka.

Summary

The research line followed in this work is the use of electronic networks for the development and deployment of new kinds of Network Distributed Decision Support Systems (NDDSSs) in the fields of water and environment. Development of these systems stems from the growing awareness that the processes of decision making in the fields of water and the environment are of a *sociotechnical* nature. The emphasis on the social domain in which the derived decisions are to be implemented is of crucial importance. Traditionally, the development of decision support systems and tools were mainly concerned with various corporate interests, mainly from industry and government. This approach tended either to marginalise the interests of some stakeholders during the decision making process, or at best to lump these interests in 'averaged' or 'standard' stakeholders. It has been recognised, however, that the actual social domain is far more complex than these assumptions allow and that without a direct participation of all the interested parties, including the general public, the decisions are hardly implementable in practice. Direct participation, on the other hand, is generally a very expensive process and very often limited by the technical means available.

The NDDSSs are envisaged as decision-making environments, deployed on the Internet, which will allow for the direct participation of all the interested parties in the process, including the general public. In particular the 'general public' is not to be treated as just one stakeholder, but as a collection of individuals each with his or her own beliefs and attitudes, which, combined with the relevant scientific data (facts), result in different judgements about the issues at hand. The processes of collaboration and negotiation proceed on the basis of these judgements, initially unstructured, but gradually leading to more complying agreements.

The first part of this study tries to provide a broad theoretical background, which is seen as necessary for understanding the sociotechnical nature of the current decision making processes. The very fact that by tradition we are used to talking about society and technology as two separate entities has its deeper origins in the most significant transformation of human thought in our history during the period known as the 'modern era'. Most of these transformations were in fact concentrated during a specific period of the modern era, identified with the project of the 'Enlightenment' and the subsequent waves of its critiques. The analysis is carried further into our present societies, which are frequently described as post-industrial societies, and our contemporary culture, commonly described as a postmodern culture. It is an era where we look with a critical eye on the project of 'modernity' which again brings about a new understanding of the nature of knowledge.

The NDDSSs themselves are envisaged in terms of this changing nature of knowledge, which may be described by two key positions. Firstly, the recognition of the fact that knowledge is increasingly produced, distributed and consumed as a *commodity,* whereby the dominant, and apparently the only legitimising criterion of any knowledge, becomes the criterion of *performativity*. Secondly, the re-discovery of the power and relevance of so-called *narrative* knowledge (sometimes the terms 'custom' knowledge, 'traditional' knowledge or 'social' knowledge are used), which during the whole project of the Enlightenment has been continuously undervalued in comparison with *scientific* knowledge. It is argued that it is precisely the neglect of narrative knowledge that so often leads to 'failures' of projects involving 'knowledge-management' components.

At the end of this first part, the analysis, design and development of NDDSSs are identified as activities most appropriately approached from within the field of *hydroinformatics*, and the systems themselves are seen as hydroinformatics systems. Hydroinformatics is defined essentially as a technology, but its social dimension has been recognised from its very beginning. In fact, the development of hydroinformatics changes its nature in a direction where it is no longer sufficient to talk only about its social dimensions, but possibly to redefine it as a sociotechnical discipline.

The second part of the study introduces more specifically the basic requirements of an NDDSS, its general concept, and its functional components. Conceptually divided into a (scientific) knowledge centre and users' periphery, an NDDSS aims at mobilising both the narrative and the scientific knowledge in the processes of decision making. The main tasks of these systems then become the transmissions of scientific knowledge from a 'centre' (which is the primary repository of scientific knowledge) to the users' periphery, and of the narrative knowledge from the users' periphery to the centre. It is proposed here that in order to achieve adequate transmission and translation of these different kinds of knowledge between the knowledge centre and the users' periphery, and their mobilisation for decision making purposes, these systems will have to be composed of three basic functional components: 1. Fact engine component, located in the knowledge centre, with the primary role of gathering, organising, providing and distributing relevant scientific knowledge as scientific facts; 2. Judgement engine component, located at the user periphery, with the primary role of combining the scientific facts with the beliefs and attitudes (traditionally represented as some sort of 'interests' and 'intentions') of the users/participants in the decision making process, in order to provide judgements of that particular user or user group about proposed plans or interventions; 3. A platform for negotiation and collaboration, where by using appropriate aggregation procedures the diverse judgements are mobilised for negotiation and collaboration purposes, which will allow for the *emergence* of commonly accepted, equitable courses of action, with appropriate compensatory arrangements. The key quality that needs to be introduced into these systems throughout the processes of knowledge transmission and translation, is the quality of *transparency*.

The third part of the thesis presents a description of a prototype of an NDDSS, which has been developed within the framework of this study. This prototype, named *Aquavoice*, has been primarily designed as a *multi-player role play* game which can be played over the Internet. *Aquavoice* has a fact engine component, a judgement engine component and some facilities for negotiation and collaboration. It has been developed for a hypothetical case study, which deals with the problems of lowering groundwater tables due to increased abstractions of groundwater for water supply. A number of conflicting interests arise represented by four different institutions, and by any number of individuals who may participate in the decision-forming process. The institutions (represented by four different role players) involved are: Local Authority, Environmental Agency, Water Supply Company, and a Farmers' Council. Each of these four institutions has different responsibilities and interests. The number of players that can take the role of concerned individuals is conceived as unlimited.

Aquavoice provides a dynamic environment for judging different scenarios of groundwater abstractions, by each user/participant, according to his or her own interests. These different scenarios can be evaluated by each user of the system, individually. The individual evaluations are then aggregated in order to present the overall preference relations of the whole community. This aggregation, together with the transparent positions of each participant in *Aquavoice* represents the basic 'social landscape' of the community and may be used to initiate the processes of negotiation and collaboration between the participating parties.

At the core of *Aquavoice* is a sophisticated hydrological model of the area in question developed by using the *MIKE SHE* modelling system. This model acts as fact engine for *Aquavoice*. Being a fully integrated physically based modelling system for simulation of the whole land phase of the hydrological cycle, *MIKE SHE* is particularly suited for being at the core of a system like *Aquavoice*. In order to deploy a *MIKE SHE* model on the Internet, several new technical developments were required. Development of a complete new user interface, adapted for the needs of *Aquavoice*, and operating in a browser environment was a first requirement. Further to this, an Internet-based communication link from this user interface with the core *MIKE SHE* model has been developed.

For the judgement engine component of *Aquavoice* some methods from the field of multi criteria analysis (MCA) have been adapted. In particular, algorithms for Fuzzy Multiple Attribute Decision Making have been implemented, in order to obtain the judgements and the evaluations of the users/participants. These methods are here used only as tools for establishing the initial response of the community to the proposed scenarios, serving as a starting point for processes of negotiation and various collaborations. The processes of collaboration and negotiations can then proceed through a chat-like interface which is provided in *Aquavoice*, and any changes of judgements or evaluations which result from these processes are immediately and transparently reflected to all the users of the system.

The main technology used for the development of the *Aquavoice* prototype is Java. This powerful language/platform technology was shown to be particularly useful for development of this kind of Internet applications.

Aquavoice has to be considered only as a small first step in the development of NDDSSs. Both in the conceptual design and in all of the three functional components of a NDDSS there are various aspects that will need further research. In the last part of the thesis the paradigm of 'agent orientation' is introduced as relevant conceptualisation alternative which can be very useful for future developments of NDDSSs. In particular, the potential of so-called 'software agent technology' for conceptualising and implementing future NDDSSs is addressed. One possible conceptualisation of an NDDSS using agent-based architecture i proposed.

The concluding part of the thesis summarises all the aspects of the development of NDDS! presented in it. It revises the need for such systems, the approaches that can be followed i their design and development, and the possible ways forward. One crucial question related t the possible ways for introduction of these systems into our societies is also addressed. Thre possible approaches for the introduction of NDDSSs into practice are proposed. The first (so called edutainment: *edu*cation+enter*tainment* approach) proposes further developments o systems similar to *Aquavoice* as environmental computer-games, in order to raise people' awareness of the possibilities of these systems. The second approach argues for th possibility of the introduction of these systems through actions of governmental and non governmental agencies, primarily at sub-national and local levels. The third approach briefl investigates the possibility of introducing these systems through the market directly. It i most likely that for different decision contexts, and in different societies and culture different

Samenvatting

(Summary in Dutch)

Dit onderzoekt richt zicht op het gebruik van elektronische netwerken voor de ontwikkeling en toepassing van nieuwe over computernetwerken verspreidde beslissingsondersteunende systemen (network distributed decision support system; NDDSS) op het gebied van water en milieu. De ontwikkeling van dit soort systemen komt voort uit het groeiend besef dat besluitvormingsprocessen op het gebied van water en milieu een sociotechnisch karakter hebben. Het is van groot belang dat de sociale context waarin de te nemen beslissingen moeten worden uitgevoerd wordt begrepen. Van oudsher hadden alleen een aantal maatschappelijke organisaties die het bedrijfsleven en de overheden vertegenwoordigen belang bij de ontwikkeling van beslissingsondersteunende systemen en instrumenten. Het gevolg van deze benadering was dat de belangen van sommige directe belanghebbenden in het besluitvormingsproces minimaal was of hoogstens samen met die van anderen werd meegenomen als een "gemiddelde" of "standaard". Het besef groeit echter dat de werkelijke maatschappelijke situatie veel ingewikkelder is dan met dergelijke aannames kan worden weergegeven en dat zonder directe deelname van alle belanghebbende partijen, waaronder ook de lokale bevolking, voorgestelde beslissingen in praktijk nauwelijks uitvoerbaar zijn. Rechtstreekse medezeggenschap houdt echter meestal een zeer kostbare procedure in en is vaak moeilijk uitvoerbaar omdat de geschikte technische middelen ontbreken.

Een NDDSS moet worden gezien als een besluitvormingsomgeving, die gebruik maakt van het internet en die het mogelijk maakt dat alle belanghebbenden, ook het brede publiek, rechtstreeks aan het besluitvormingsproces deelnemen. Vooral het "brede publiek" kan niet worden beschouwd als één enkele belanghebbende, maar moet worden gezien als een verzameling individuen met elk eigen overtuiging en waarheden. Dit leidt, tezamen met relevante wetenschappelijke gegevens (feiten), tot andere beoordelingen over de kwesties die aan de orde zijn. De processen van samenwerking en onderhandeling verlopen op basis van deze beoordelingen, en leiden tot aanvankelijk ongestructureerde, maar geleidelijk tot meer gedragen afspraken.

In het eerste deel van dit onderzoek wordt een algemene theoretische achtergrond geschetst, die noodzakelijk is om het sociotechnische karakter van het bestaande besluitvormingsproces te begrijpen. Het feit dat we gewend zijn maatschappij en technologie als twee aparte werelden te beschouwen heeft zijn oorsprong in de belangrijkste ommekeer in het menselijk denken in onze geschiedenis, die plaatsvond in de periode die we betitelen als de "Moderne tijd". Deze ommekeer vond in feite plaats tijdens een korte periode van de moderne tijd, door een stroming die de "Verlichting" wordt genoemd en de daaropvolgende kritische reacties. Deze denkbeelden werken door in onze huidige maatschappij, die vaak wordt beschreven als de postindustriële maatschappij, en in onze hedendaagse cultuur, die algemeen wordt beschreven als postmodernistische cultuur. Heden ten dage wordt het "modernisme" opnieuw aan een kritische beschouwing onderworpen, hetgeen leidt tot een nieuw begrip van de aard van kennis.

Een NDDSS moeten worden gezien in het licht van deze wisselende waardering van kennis, die in twee sleutelbegrippen kan worden samengevat. Op de eerste plaats komt de erkenning van het feit dat er steeds meer kennis beschikbaar komt, uitgedragen wordt en gebruikt als handelsartikel, waarbij de belangrijkste en blijkbaar enige legitieme maatstaf van elke soort kennis, *vervulbaarheid* (*'performativity'*) is. Op de tweede plaats komt de herontdekking van de kracht en relevantie van zogenaamde *verhalende* (*'narrative'*) kennis (ook wel bekend als "huis-tuin-en-keuken" kennis, "boerenverstand", "traditionele" kennis of "maatschappelijke" kennis). Dit soort kennis wordt sinds de Verlichting systematisch ondergewaardeerd in vergelijking met *wetenschappelijke* kennis. Er wordt wel beweerd dat projecten waarin kennis- en informatiesystemen een rol spelen zo vaak mislukken, juist omdat die verhalende kennis wordt verwaarloosd.

Aan het einde van dit eerste deel worden analyse, ontwerp en ontwikkeling van NDDSSen beschreven als activiteiten die het best kunnen worden benaderd vanuit het gezichtspunt van *hydro-informatica* ("hydroinformatics") en worden de systemen zelf als hydro-informatica systemen gezien. Hydro-informatica wordt in feite als een technologie beschouwd, maar al vanaf het allereerste begin heeft men de maatschappelijke dimensie ervan onderkend. In feite is de aard van hydro-informatica veranderd in de loop van de ontwikkeling ervan zodat er niet meer alleen kan worden gesproken van de bijbehorende maatschappelijke dimensie, maar dat het vakgebied waarschijnlijk tot de sociotechnische disciplines moet worden gerekend.

In het tweede deel van dit onderzoek worden de basiseisen, algemene begrippen en functionele componenten van een NDDSS geïntroduceerd. Conceptueel gescheiden in een (wetenschappelijk) kenniscentrum met daaromheen een gebruikersomgeving, is het doel van een NDDSS bij het besluitvormingsproces gebruik te maken van zowel verhalende als formele wetenschappelijke kennis. De voornaamste taak van deze systemen wordt zo het overdragen van wetenschappelijke kennis vanuit een "centrum" (waar wetenschappelijke kennis primair is opgeslagen), naar de gebruikersomgeving en van het overdragen van verhalende kennis uit de gebruikersomgevingen naar het centrum. We stellen ons voor dat er tussen het kenniscentrum en de gebruikersomgevingen een adequate overdracht en vertaling plaatsvindt van deze verschillende soorten kennis, zodat beide kunnen worden gebruikt bij de besluitvorming. Daartoe dienen deze systemen drie functionele basiscomponenten te bevatten: 1. een feitenverwerkende component ('fact-engine'), die zich bevindt in het kenniscentrum en waarvan de voornaamste functie is het verzamelen, organiseren, beschikbaar stellen en distribueren van relevante wetenschappelijke kennis in de vorm van wetenschappelijke feiten; 2. een beoordelende component ('judgement-engine'), die zich bevindt in de gebruikersomgeving en waarvan de voornaamste functie is het combineren van de wetenschappelijke gegevens met de overtuigingen en waarheden (traditioneel beschouwd als "belangen" en "bedoelingen") van gebruikers van- of deelnemers aan het besluitvormingsproces, zodat deze gebruikers of gebruikersgroepen een gefundeerd besluit kunnen nemen over voorgestelde plannen en voornemens; 3. een overleg- en samenwerkingsplatform, waarin met gebruik van de juiste koppelingsprocedures verschillende beoordelingen worden gemobiliseerd, zodat ze in het overleg- en samenwerkingsproces kunnen worden gebruikt. Hieruit zal dan een algemeen geaccepteerde, evenwichtige koers *voortvloeien*, met passende compensatieregelingen. De voornaamste voorwaarde waaraan alle kennisoverdracht- en kennisvertalingsprocessen tussen deze systemen moeten voldoen is *helderheid.*

Het derde deel van dit proefschrift beschrijft een prototype van een NDDSS, dat speciaal voor dit onderzoek is ontwikkeld. Dit prototype, genaamd *Aquavoice*, is vooral ontwikkeld als rollenspel voor meer spelers dat gespeeld kan worden via het Internet. *Aquavoice* heeft een fact-engine component, een judgement-engine component en biedt mogelijkheden tot onderhandelen en samenwerken. Het programma is ontwikkeld voor een hypothetische situatie, waarin problemen worden bekeken die te wijten zijn aan een daling van de grondwaterspiegel in verband met de toename van grondwaterwinning. Het blijkt dat vier verschillende organisaties en een aantal personen die deelnemen aan het besluitvormingsproces tegengestelde belangen hebben. De organisaties in kwestie (de rol daarvan wordt gepeeld door vier verschillende spelers) zijn: de plaatselijke overheid, de milieuorganisatie, het drinkwaterbedrijf en een boerenorganisatie. Al deze vier organisaties hebben verschillende verantwoordelijkheden en belangen. Het aantal spelers dat de rol van geëngageerde burgers kan vertolken is onbeperkt.

Aquavoice biedt elke gebruiker/deelnemer een dynamische omgeving om verschillende scenario's voor grondwaterwinning te beoordelen, toegespitst op diens eigen specifieke belangen. Deze verschillende scenario's kunnen door elke gebruiker van het systeem afzonderlijk worden geëvalueerd. Deze individuele evaluaties worden vervolgens samengevoegd om de overkoepelende voorkeurrelaties van de hele gemeenschap te laten zien. Deze samenvoeging en de heldere posities van de verschillende deelnemers aan *Aquavoice* vertegenwoordigen het "maatschappelijke landschap" van de gemeenschap hetgeen kan worden gebruikt om het proces van onderhandeling en samenwerking tussen de verschillende deelnemers op gang te brengen.

Het hart van *Aquavoice* is een geavanceerd hydrologisch model van het betrokken gebied dat ontwikkeld is met behulp van het *MIKE SHE* modelleersysteem. Dit model dient als fact-engine voor *Aquavoice*. Als volledig geïntegreerd ruimtelijk modelleersysteem voor het algehele landelijke deel van de hydrologische cyclus, is *MIKE SHE* is bijzonder geschikt om te dienen als hart van een systeem als *Aquavoice*. Om een *MIKE SHE* model geschikt te maken voor gebruik op Internet waren verschillende nieuwe technieken nodig. Er moest een totaal nieuwe gebruikersinterface worden ontwikkeld die toegespitst is op de behoeftes van *Aquavoice*, en het was essentieel dat dit in een browser-omgeving werkte. Bovendien werd een op Internet-gebaseerde verbinding ontwikkeld om de gebruikersinterface aan de *MIKE SHE*-kern te koppelen.

Voor het judgement-engine gedeelte van *Aquavoice* werden aangepaste technieken op het gebied van Multi Criteria Analysis (MCA) gebruikt. In het bijzonder werden algoritmes voor Fuzzy Multiple Attribute Decision Making toegepast, om het oordeel en de evaluaties van de gebruikers/deelnemers te verkrijgen. Deze technieken zijn hier slechts als instrument gebruikt om de eerste reactie van een gemeenschap op de voorgestelde scenario's te peilen en dienen als uitgangspunt voor verschillende onderhandelings- en samenwerkingsrondes. De onderhandelings- en samenwerkingsrondes worden daarna vervolgd in een chat-boxachtige omgeving, aangeboden in *Aquavoice*, en alle veranderingen in beoordeling en evaluatie die hieruit voortvloeien worden onmiddellijk duidelijk zichtbaar voor alle gebruikers van het systeem.

De voornaamste technologie die gebruikt is voor het ontwikkelen van het *Aquavoice* prototype is Java. Deze krachtige taal/platform technologie bleek bijzonder geschikt voor het ontwikkelen van dit soort Internet-toepassingen.

Aquavoice moet worden beschouwd als eerste stap in de richting van de ontwikkeling van een NDDSS. Zowel wat betreft het conceptuele ontwerp als wat betreft alle drie de functionele componenten van een NDDSS blijven er verschillende aspecten over die nader moeten worden uitgewerkt. In het laatste deel van dit proefschrift wordt het paradigma van 'agent orientation' geïntroduceerd als een conceptueel alternatief dat voor toekomstige ontwikkelingen op NDDSS-gebied erg nuttig kan zijn. Vooral de mogelijkheden van zogenaamde 'software-agent technology' voor het conceptualiseren en realiseren van toekomstige NDDSSen worden behandeld. Een mogelijke conceptualisatie van een NDDSS waarbij agent-technieken worden gebruikt wordt voorgesteld.

In het afsluitende deel van dit proefschrift worden alle aspecten van de ontwikkeling van een NDDSS nader beschouwd. Het bevat een samenvatting van de behoefte aan dit soort systemen, de mogelijke benaderingen van het opzetten en ontwikkelen van NDDSSen en mogelijke toekomstige ontwikkelingen. Een essentieel aspect, dat te maken heeft met mogelijke manieren van het invoeren van zo'n systeem in onze maatschappij, wordt ook behandeld. Er zijn drie mogelijke manieren om de invoering van NDDSSen in praktijk aan te pakken. Een mogelijke eerste methode van aanpak is het zogenaamde edutainment: *edu*catie+enter*tainment*. Zo zouden toekomstige ontwikkelingen van op *Aquavoice* gelijkende systemen kunnen worden toegepast in computerspelletjes waardoor mensen zich bewust worden van de mogelijkheden van dit soort systemen. Een mogelijke tweede methode van aanpak is dat overheids- en andere instanties de invoering van dit soort systemen actief aanmoedigen, vooral op lokaal en provinciaal niveau. Een mogelijke derde methode van aanpak wordt kort aangestipt: de mogelijkheid dit soort systemen rechtstreeks op de markt te brengen. Waarschijnlijk zullen in verschillende besluitvormingsomstandigheden en in andere samenlevingen en culturen verschillende methoden van aanpak moeten worden gebruikt.

About the author

Andreja Jonoski was born in 1963 in Kumanovo, Republic of Macedonia. He graduated from the Faculty of Civil Engineering at the Sts. Cyril and Methodius University of Skopje, Macedonia, in 1988. Until 1994 he worked at the Civil Engineering Institute "Macedonia", in its geotechnics division, first as a groundwater hydrologist, and later as head of the division. He was involved in a number of projects related to groundwater hydrology, hydrogeology, engineering geology, soil mechanics and geotechnics.

In 1994 he joined the International Institute for Infrastructural, Hydraulic, and Environmental Engineering (IHE), in Delft, The Netherlands, as a participant in the international postgraduate programme in groundwater hydrology. He obtained his Postgraduate Diploma in Groundwater Hydrology with distinction in 1995, and, one year later, in 1996, he obtained his MSc degree, again with distinction, with a study on coupling groundwater simulation models and optimisation models for the improved design of artificial recharge systems in the Netherlands.

The research presented in this study became his main preoccupation in 1996 when he became enrolled in the PhD programme of IHE as a Research Fellow in the hydroinformatics section. His main research interests are design and development of Internet- and Web-based decision support systems for water- and environment-related problems, as well as hydrological and groundwater modelling. He has published several papers in international journals and conference proceedings. Since 2001 he has worked at IHE as a lecturer in the international postgraduate programme in hydroinformatics.